AF361923

Global Vegetation and Land Surface Dynamics in a Changing Climate

Global Vegetation and Land Surface Dynamics in a Changing Climate

Editors

Pinki Mondal
Sonali Shukla McDermid

MDPI • Basel • Beijing • Wuhan • Barcelona • Belgrade • Manchester • Tokyo • Cluj • Tianjin

Editors
Pinki Mondal   Sonali Shukla McDermid
University of Delaware   New York University
USA   USA

Editorial Office
MDPI
St. Alban-Anlage 66
4052 Basel, Switzerland

This is a reprint of articles from the Special Issue published online in the open access journal *Land* (ISSN 2073-445X) (available at: https://www.mdpi.com/journal/land/special_issues/gloveg).

For citation purposes, cite each article independently as indicated on the article page online and as indicated below:

LastName, A.A.; LastName, B.B.; LastName, C.C. Article Title. *Journal Name* **Year**, *Volume Number*, Page Range.

ISBN 978-3-0365-0502-2 (Hbk)
ISBN 978-3-0365-0503-9 (PDF)

Cover image courtesy of Pinki Mondal.

Contents

About the Editors

Pinki Mondal is a geospatial data scientist interested in the dynamics of coupled natural and human systems. She is an Assistant Professor in the Department of Geography and Spatial Sciences at the University of Delaware, USA, and a Resident Faculty at the UD Data Science Institute. With expertise in environmental remote sensing and Geographic Information Systems (GIS), she examines the impacts of climate change on diverse ecosystems around the world. Dr. Mondal has a Ph.D. in Land Change Science from the University of Florida. Prior to joining the University of Delaware, she was a Senior Research Associate at Columbia University, USA.

Sonali Shukla McDermid is a climate scientist and Associate Professor of Environmental Studies at NYU. Her research, which uses global climate models, crop models, and observational datasets, focuses on understanding how agricultural land management has transformed our climate and regional environments. She has also served as Climate Co-Lead for the Agricultural Intercomparison and Improvement Project (www.agmip.org), which assessed the impact of climate change on food security and livelihoods across South Asia and Sub-Saharan Africa. McDermid holds a B.A. in Physics from NYU (2006) and an M.Phil and Ph.D. (2012) from the Dept. of Earth and Environmental Sciences at Columbia University in Atmospheric Science and Climatology. Prior to NYU, she was NASA Post-Doctoral Fellow at the Goddard Institute for Space Studies in New York City.

Editorial

Editorial for Special Issue: "Global Vegetation and Land Surface Dynamics in a Changing Climate"

Pinki Mondal [1,2,*] **and Sonali Shukla McDermid** [3]

[1]   Department of Geography and Spatial Sciences, University of Delaware, Newark, DE 19716, USA
[2]   Department of Plant and Soil Sciences, University of Delaware, Newark, DE 19716, USA
[3]   Department of Environmental Studies, New York University, New York, NY 10003, USA; sps246@nyu.edu
*   Correspondence: mondalp@udel.edu

Received: 29 December 2020; Accepted: 4 January 2021; Published: 6 January 2021

Global ecosystem changes have multiple drivers, including both natural variability and anthropogenic climate and environmental change. Intensifying climate change, inclusive of increased hydroclimate variability and extremes, temperatures, CO_2 fertilization effects, and even changes in cloud cover can produce competing vegetation responses, changes in underlying soil and land surface health, and drive overall changes in ecosystem productivity [1–3]. Anthropogenic pressures also extend beyond climate change to include the clearing of native forests and other vegetation for fuel, fiber, food, increasing infrastructure and municipal development, as well as the intensive management and expansion of agricultural lands and soils for crop production, which can result in vegetation and ecosystem degradation and/or wholesale losses.

Furthermore, these individual climate and environmental drivers incite complex vegetation and ecosystem interactions and feedbacks that can exacerbate each other and/or produce competing effects that make identifying robust ecosystem changes challenging [4]. As one example, atmospheric CO_2 fertilization effects are known to decrease plant water demand via reductions in stomatal conductance [5,6]. However, enhanced atmospheric CO_2 can also increase leaf area index (LAI), thereby increasing total evapotranspiration (ET) and offsetting the stomatal effects [7,8]. While this can cool the surface, particularly in arid to semi-arid regions [9], enhanced ET may also exacerbate soil moisture declines driven by extended dry seasons, reduced snowpack and more variable rainfall. Such feedbacks may imperil greening trends, particularly in more water-limited regions, under future climate and land use trajectories. To resolve these interactions, we must move beyond observations of "greening" and "browning" (a decline in ecosystem productivity) alone towards more integrated climate and vegetation information. It is thus important that ecosystem monitoring and evaluation incorporate measures of the relative presence and influence of these feedbacks, in addition to tracking changes in vegetation and greening/browning.

Of the numerous climate and environmental drivers, a substantial fraction of global greening has been attributed to CO_2 fertilization effects, with regional climate changes also driven partly by nutrient deposition, and land management and change [2,10,11]. Nevertheless, there still remain outstanding uncertainties in the drivers of greening vs. browning at regional and local scales. These uncertainties stem from limitations in both the observational products, such as to identify specific vegetation types and transitions, and in ecosystem models' ability to represent complex vegetation processes and competing biophysical responses, such as mortality and disturbances, changes in land use change and management, time-varying vegetation changes under different forcing trajectories [7,12,13] and other biotic factors like pests and diseases. Future climate changes and intensifying land management will bring additional, interactive changes to natural and managed ecosystems. Managing these ecosystems for current and future changes requires an improved understanding of these multiscale drivers of vegetation dynamics, the mechanisms by which they operate, and how they change over time. This special issue in Land draws together a collection of five articles covering geographic regions in Africa,

Asia and North America [14–18]. Taken together, these articles encompass a wide range of study spanning the biophysical drivers (e.g., hydrological) of vegetation change and dynamics and the human impacts (e.g., perceptions of climate change effects) in both natural and managed ecosystems.

Ugbaje and Bishop [14] examine the relative importance of hydrological controls such as soil moisture, precipitation and terrestrial water storage in determining vegetation greenness in Africa. Using a multivariate analysis, the authors report that precipitation and soil moisture are the most important predictors of vegetation greenness. However, anomalies in vegetation greenness were best predicted by the anomalies in soil moisture and terrestrial water storage in most of Africa with a diminished role of precipitation. The authors also note that predominantly positive trends in vegetation greenness anomalies might be a result of factors other than water availability, such as atmospheric fertilization, use of high yielding crop varieties, afforestation and an increase in growing season length [14].

Globally, an encroachment of woody species into savanna biomes potentially threatens important ecosystem attributes and functionality for both natural processes and human use. However, much uncertainty remains in identifying key drivers of woody encroachment and their interactions with key species, as well as in developing analytical techniques conducive to scaling these dynamics across species and the larger ecosystem. Meyer et al. [15] focus on the Kalahari landscape in Botswana as a test case to understand the environmental drivers determining woody species composition and structure. Their work moves beyond single-species analyses to consider morphological groups, which enable their findings to potentially be scaled across the greater ecosystem. They employ statistical analysis techniques that are more suitable to the distributions that characterize ecological data, making them more informative to land managers. Based on species identified, diversity and abundance data collected at multiple transects, the authors find that precipitation largely explains species richness and abundance when all morphological groups were considered together. However, precipitation was often mediated by other key factors, such that low grazing and fire also contributed to higher species richness. In contrast to previous studies, they also find that higher grazing, as indicated by increased borehole density and cattle numbers, can lead to reduced woody species abundance due to more frequent rotation of cattle and thus less soil disturbance [15].

Comer et al. [16] utilize an integrated exposure-sensitivity-adaptive capacity framework to examine climate change vulnerability of 52 major vegetation types in the western United States. The proposed framework aims to highlight the interactions between climate-induced stress and other ecological stressors. Such interactions might force natural plant communities to transform in unprecedented ways, due to their already reduced resilience. With a reproducible and transparent Habitat Climate Change Vulnerability Index (HCCVI), the authors provide an early warning system of elevated risks for natural plant species, especially valuable for the conservation practitioners. The authors find that currently 50 out of 52 vegetation types have at least moderate vulnerability, while no vegetation type exhibits very high vulnerability. Yet, when mid-21st century climate exposure projections are considered, all but 19 vegetation types shift to the high vulnerability category. They conclude that by measuring relative severity, this framework would be of particular use for gauging risk of environmental degradation for the next several decades [16].

Ongoing changes in weather parameters, such as precipitation and temperature, are known to impact crop yield in different ways. It is also well-documented that adaptation strategies in response to such changes can limit negative impacts on crop yield. Rondhi et al. [17] examine the role of climate change impact perception on farmer adaptation practices by analyzing household survey data from 87,330 farmers in Indonesia. The authors use an ordered probit regression model to analyze the effects of 17 variables from a range of factors: economic, technical, institutional and climatic. All climatic variables have a statistically significant positive impact on the farmers' perceived impact, with floods being identified as the most damaging climatic event, followed by drought, heavy rain and other hazards such as landslides. While there is little evidence to indicate any difference between actual impact and perceived impact, it is important to characterize farmers' perceived impact and factors affecting it. An exaggerated perceived impact might lead to maladaptive outcomes, such as excessive application of chemical fertilizer, which has implications for water and soil quality, and ultimately crop health [17].

Improving characterizations of ecosystem responses to drought in East Africa is both urgent, due to intensifying anthropogenic climate change and land use pressures, and challenging, owing to limitations in ground-based measurement and observations. The emergence and proliferation of satellite-based products can offer a variety of ways to monitor ecosystem change in the highly-variable East African climate, but there is a need to better understand how these products capture important ecosystem stress signals. To address this need, Robinson et al. [18] analyze how the normalized difference vegetation index (NDVI) and the solar-induced chlorophyll fluorescence (SIF) capture one of the more exceptional, recent drought periods in East Africa: the failure of both the 2010 "short" and 2011 "long" rainy seasons. They further provide a mechanistic characterization of the drought's temporal evolution, contextualizing it within regional anthropogenic climate trends. They find that despite constraints on its spatial resolution and lower energy, SIF does indeed capture vegetation drought stress similar to NDVI, both instantaneously and at a lag, which gives confidence in its use as an early indicator of regional drought conditions. Nevertheless, they stress that instruments designed with the intention of capturing drought-induced ecosystem stress are critical to the improvement and enhanced utility of SIF measurements. Robinson et al.'s analysis also highlights how reduced rainfall in one season, in this case the 2010 short rains, can compound the effects of regional drought during the subsequent long rains, exacerbating water-stress impacts to both natural and human ecosystems. This is an important temporal dynamic to better understand and monitor, as the Indian Ocean continues to warm and interact with other oceanic and atmospheric variability and change, thereby changing East African regional atmospheric circulation patterns and moisture convergence [18].

The studies included in this special issue represent a diversity of global regions, scales of analysis and analytical techniques, and highlight perspectives on both human and natural land system change. Nevertheless, some common and key messages can be distilled from these works. Firstly, while some ecosystem drivers, e.g., hydroclimate variability, emerge as being of primary importance, they are also mediated by several other factors such as fire [15], and even gender [17]. Such interactions underscore the complexity and dynamism of both natural and human ecosystem responses to global environmental change, which must increasingly be accounted for in quantitative analysis techniques. In addition, several of these studies highlight that key drivers of ecosystem change may themselves be subject to time-varying trends that potentially influence their magnitude, timing, and/or interactions with other drivers or different ecosystem components. Furthermore, all of these studies highlight the importance of improved research, by way of data collection and analysis techniques, instrumentation (e.g., satellite sensor) development, and indicators and metrics for future evaluations of global and regional ecosystem change. These advances are critical to help inform and guide not only novel research, but also more effective decision-making for landscape and ecosystem managers in an era of increasing anthropogenic pressures.

Author Contributions: Conceptualization, P.M. and S.S.M.; writing—original draft preparation, P.M. and S.S.M.; writing—review and editing, P.M. and S.S.M. All authors have read and agreed to the published version of the manuscript.

Funding: This research received no external funding.

Acknowledgments: We thank all the reviewers for their feedback on earlier versions of the manuscripts in this Special Issue.

Conflicts of Interest: The authors declare no conflict of interest.

References

1. Notaro, M.; Vavrus, S.; Liu, Z. Global vegetation and climate change due to future increases in CO_2 as projected by a fully coupled model with dynamic vegetation. *J. Clim.* **2007**, *20*, 70–90. [CrossRef]

2. Zhu, Z.; Piao, S.; Myneni, R.B.; Huang, M.; Zeng, Z.; Canadell, J.G.; Ciais, P.; Sitch, S.; Friedlingstein, P.; Arneth, A.; et al. Greening of the Earth and its drivers. *Nat. Clim. Chang.* **2016**, *6*, 791–795. [CrossRef]

3. D'odorico, P.; Bhattachan, A.; Davis, K.F.; Ravi, S.; Runyan, C.W. Global desertification: Drivers and feedbacks. *Adv. Water Resour.* **2013**, *51*, 326–344. [CrossRef]

4. Bajželj, B.; Richards, K. The Positive Feedback Loop between the Impacts of Climate Change and Agricultural Expansion and Relocation. *Land* **2014**, *3*, 898–916. [CrossRef]

5. Donohue, R.J.; Roderick, M.L.; McVicar, T.R.; Farquhar, G.D. Impact of CO_2 fertilization on maximum foliage cover across the globe's warm, arid environments. *Geophys. Res. Lett.* **2013**, *40*, 3031–3035. [CrossRef]

6. Higgins, S.I.; Scheiter, S. Atmospheric CO_2 forces abrupt vegetation shifts locally, but not globally. *Nature* **2012**, *488*, 209–212. [CrossRef] [PubMed]

7. Mankin, J.S.; Smerdon, J.E.; Cook, B.I.; Williams, A.P.; Seager, R. The Curious Case of Projected Twenty-First-Century Drying but Greening in the American West. *J. Clim.* **2017**, *30*, 8689–8710. [CrossRef] [PubMed]

8. Zhang, K.; Kimball, J.S.; Nemani, R.R.; Running, S.W.; Hong, Y.; Gourley, J.J.; Yu, Z. Vegetation Greening and Climate Change Promote Multidecadal Rises of Global Land Evapotranspiration. *Sci. Rep.* **2015**, *5*, 15956. [CrossRef] [PubMed]

9. Forzieri, G.; Alkama, R.; Miralles, D.G.; Cescatti, A. Satellites reveal contrasting responses of regional climate to the widespread greening of Earth. *Science* **2017**, *356*, 1180–1184. [CrossRef] [PubMed]

10. Mao, J.; Ribes, A.; Yan, B.; Shi, X.; Thornton, P.E.; Séférian, R.; Ciais, P.; Myneni, R.B.; Douville, H.; Piao, S.; et al. Human-induced greening of the northern extratropical land surface. *Nat. Clim. Chang.* **2016**, *6*, 959–963. [CrossRef]

11. Chen, C.; Park, T.; Wang, X.; Piao, S.; Xu, B.; Chaturvedi, R.K.; Fuchs, R.; Brovkin, V.; Ciais, P.; Fensholt, R.; et al. China and India lead in greening of the world through land-use management. *Nat. Sustain.* **2019**, *2*, 122–129. [CrossRef] [PubMed]

12. Levis, S. Modeling vegetation and land use in models of the Earth System. Wiley Interdiscip. *Rev. Clim. Chang.* **2010**, *1*, 840–856. [CrossRef]

13. Mahowald, N.; Lo, F.; Zheng, Y.; Harrison, L.; Funk, C.; Lombardozzi, D.; Goodale, C. Projections of leaf area index in earth system models. *Earth Syst. Dyn.* **2016**, *7*, 211–229. [CrossRef]

14. Ugbaje, S.U.; Bishop, T.F.A. Hydrological control of vegetation greenness dynamics in Africa: A multivariate analysis using satellite observed soil moisture, terrestrialwater storage and precipitation. *Land* **2020**, *9*, 15. [CrossRef]

15. Meyer, T.; Holloway, P.; Christiansen, T.B.; Miller, J.A.; D'Odorico, P.; Okin, G.S. An assessment of multiple drivers determining woody species composition and structure: A case study from the Kalahari, Botswana. *Land* **2019**, *8*, 122. [CrossRef]

16. Comer, P.J.; Hak, J.C.; Reid, M.S.; Auer, S.L.; Schulz, K.A.; Hamilton, H.H.; Smyth, R.L.; Kling, M.M. Habitat climate change vulnerability index applied to major vegetation types of thewestern interior United States. *Land* **2019**, *8*, 108. [CrossRef]

17. Rondhi, M.; Khasan, A.F.; Mori, Y.; Kondo, T. Assessing the role of the perceived impact of climate change on national adaptation policy: The case of rice farming in Indonesia. *Land* **2019**, *8*, 81. [CrossRef]

18. Robinson, E.S.; Yang, X.; Lee, J.-E. Ecosystem Productivity and Water Stress in Tropical East Africa: A Case Study of the 2010–2011 Drought. *Land* **2019**, *8*, 52. [CrossRef]

Publisher's Note: MDPI stays neutral with regard to jurisdictional claims in published maps and institutional affiliations.

 *land*

Article

Hydrological Control of Vegetation Greenness Dynamics in Africa: A Multivariate Analysis Using Satellite Observed Soil Moisture, Terrestrial Water Storage and Precipitation

Sabastine Ugbemuna Ugbaje [1],* and Thomas F.A. Bishop [2]

[1] School of Life and Environmental Sciences, The University of Sydney, Sydney NSW 2006, Australia
[2] Sydney Institute of Agriculture & School of Life and Environmental Sciences, The University of Sydney, Sydney NSW 2006, Australia; thomas.bishop@sydney.edu.au
* Correspondence: sabastine.ugbaje@sydney.edu.au

Received: 1 December 2019; Accepted: 7 January 2020; Published: 10 January 2020

Abstract: Vegetation activity in many parts of Africa is constrained by dynamics in the hydrologic cycle. Using satellite products, the relative importance of soil moisture, rainfall, and terrestrial water storage (TWS) on vegetation greenness seasonality and anomaly over Africa were assessed for the period between 2003 and 2015. The possible delayed response of vegetation to water availability was considered by including 0–6 and 12 months of the hydrological variables lagged in time prior to the vegetation greenness observations. Except in the drylands, the relationship between vegetation greenness seasonality and the hydrological measures was generally strong across Africa. Contrarily, anomalies in vegetation greenness were generally less coupled to anomalies in water availability, except in some parts of eastern and southern Africa where a moderate relationship was evident. Soil moisture was the most important variable driving vegetation greenness in more than 50% of the areas studied, followed by rainfall when seasonality was considered, and by TWS when the monthly anomalies were used. Soil moisture and TWS were generally concurrent or lagged vegetation by 1 month, whereas precipitation lagged vegetation by 1–2 months. Overall, the results underscore the pre-eminence of soil moisture as an indicator of vegetation greenness among satellite measured hydrological variables.

Keywords: vegetation activity; vegetation anomaly; random forest

---

1. Introduction

The hydrologic component of the climate system, as a vital driver of vegetation activity and productivity across many terrestrial ecosystems, is a constraint to over half of the world's primary productivity [1]. Consequently, shifts and anomalies in the dynamics of the hydrologic cycle have far-reaching impacts not only on vegetation but also on human livelihood and wildlife. There is, therefore, the need for consistent monitoring of the hydrologic cycle in tandem with vegetation activity, which is critical to the understanding of the influence of climate variability and change on natural and agricultural systems. This is also important for early warning systems and attaining sustainable use of water resources.

Recent advances in remote sensing, especially satellite tracking systems, have enabled the consistent monitoring of vegetation dynamics. Thus, satellite-retrieved data on vegetation and components of the hydrologic cycle is often used to probe the degree to which water availability is coupled to vegetation dynamics. While the majority of these studies are focused on unraveling the relationship between vegetation and precipitation (e.g., [2,3]), a few others compare the strength of the relationship between vegetation-precipitation and vegetation-soil moisture, e.g., [4]. This comparison has recently been extended to vegetation-precipitation versus vegetation-terrestrial water storage (TWS) from the Gravity Recovery and Climate Experiment (GRACE) satellites, e.g., [5,6]. However, the challenges with these studies are (i) the relative importance of the variables is drawn from multi-temporal bivariate correlation analyses wherein the time series of vegetation and one of the hydrological variables at various lags are assessed one at a time, rather than from a multivariate analysis; (ii) prior to the analysis, seasonality in the time series is often removed by subtracting the long-term monthly climatology from the corresponding monthly observations across the years. The resulting time series is termed anomalies. The motivation for this is, in part, to allow the use of parametric statistics in which the assumptions of stationarity and independence might be violated by the presence of seasonality and autocorrelation in the time series. However, an analysis based on anomalies alone will mask the impact of the hydrological measures on vegetation phenology in ecosystems with markedly wet and dry season oscillations. All in all, these studies often assume a linear relationship between vegetation greenness and the hydrological variables, thereby overlooking the potential for interactions between the hydrological measures and their lag effects on vegetation dynamics. To overcome this challenge would require the use of a more robust multivariate approach in which simultaneous analysis and untangling of the relative importance of the hydrological variables and their lags are carried out, regardless of whether the original observation or anomalies are used.

Given these gaps, the aim of this study is to perform a multivariate analysis using time series of vegetation greenness as response variable and precipitation, soil moisture, and TWS at eight concurrent or lead lags as predictor variables. The specific research questions associated with the objective of the study are

(1) What was the spatiotemporal trend in vegetation greenness across Africa from 2003 to 2015?

(2) How well is the dynamics in vegetation greenness associated with the combined influence of precipitation, soil moisture, and TWS, and what are their relative contributions?

The analysis was performed using either the original or anomalies of the monthly values of the variables, and the model was estimated with the random forest algorithm. Random forest is a non-parametric, distribution-free machine learning algorithm that is capable of modeling linear and non-linear relationships between variables [7]. The study is focused on Africa, which is a continent with considerable water-limited ecosystems [8,9] that severely affect livelihoods. In addition, many studies that included soil moisture in assessing the eco-hydrological relationship between water availability and vegetation greenness dynamics in Africa largely used modeled soil information [9–11], essentially because of the paucity of in situ data. Thus, the use of independent satellite-observed datasets in this study should offer new insights into the eco-hydrological relationships across Africa and ameliorate the data paucity situation.

2. Materials and Methods

2.1. Study Area and Data

The study area encompasses the vegetative region of Africa, as shown by the land cover map in Figure 1. Monthly time series of the Enhanced Vegetation Index (EVI), precipitation, soil moisture, and GRACE TWS were used in this study (Table 1). Because GRACE TWS was first available in 2002 and we are interested in vegetation delayed response of up to 12 months to water availability (see Section 2.3), the data span of available hydrological variables that was considered was from 2002 to 2015, whereas EVI was from 2003 to 2015. Thus, the latter period determined the temporal extent of the analysis.

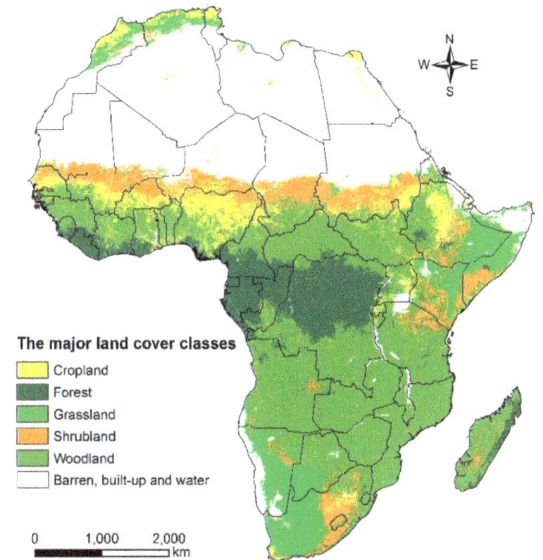

Figure 1. The major land cover types across Africa aggregated from the University of Maryland (UMD) MODIS land cover layer for 2013 (adapted from Ugbaje et al. [12]).

Table 1. General description of the research data.

	Data	Resolution	Source/Citation
EVI	MODIS Collection 6 monthly EVI Product-MOD13C2: 2002–2015	~0.05°	National Aeronautics and Space Administration's Earthdata portal (ftp://ladsweb.nascom.nasa.gov, last accessed July 2017). Didan [13]
Precipitation	CHIRPS Version 2 monthly precipitation data: 2002–2015	~5 km	Climate Hazards Group InfraRed Precipitation with Station data (CHIRPS) portal (http://chg.geog.ucsb.edu/data/chirps, last accessed July 2017). Funk et al. [14]
Soil moisture	ESA CCI daily soil moisture	~0.25°	European Space Agency Climate Change Initiative data portal (http://www.esa-soilmoisture-cci.org, last assessed July 2017). EODC [15]
Terrestrial Water Storage Anomaly (TWSA)	GRACE monthly TWSA	~1°	National Aeronautics and Space Administration's GRACE data portal (http://grace.jpl.nasa.gov, last accessed July 2017). Swenson and Wahr [16]; Landerer and Swenson [17].

EVI, as a surrogate for vegetation greenness, has the advantages of being robust against background and atmospheric noises, and, unlike the Normalized Difference Vegetation Index, it does not saturate over high biomass regions [18]. For this study, EVI time series from the MOD13C2 (version 6) product, which is a derivative of image acquisitions from the moderate resolution imaging spectroradiometer (MODIS) sensor onboard the Terra satellite, was retrieved. MOD13C2 was derived from the MOD13A2 product, which is a 16-day composite at a 1-km spatial resolution.

Monthly precipitation data were obtained from the Climate Hazards Group InfraRed Precipitation with Station data portal. CHIRPS are global, gridded precipitation datasets, derived from a blend of satellite-observed infrared cold cloud duration and in situ gauge records [14]. The blending is performed using a series of algorithms and interpolation techniques, which are described in Funk et al. [14]. Validation of CHIRPS estimates against independent ground observations and some global gridded precipitation products show CHIRPS to have a relatively low bias [14]. Further, with a spatial resolution of 0.05°, CHIRPS has one of the finest spatial resolution of all the currently available long-term, global gridded precipitation products. The products, as an integral component of the Famine Early Warning Systems Network, have been used for other applications as well, e.g., [12].

Daily satellite observed soil moisture data were downloaded from the European Space Agency Climate Change Initiative data portal. The merged product comprising retrievals from the active and passive microwave soil moisture was used. This product represents the surface soil moisture not deeper than 10 cm [19]. While the merging approach is described in Liu et al. [19,20] and Wagner et al. [21] and the global validation with ground measurements is reported in Dorigo et al. [22], the merging approach used in version 03.2 (this was used for this study) includes a weighted averaging technique where the weights are proportional to the signal-to-noise ratio [15]. As this study is at a monthly time step, we averaged the original daily observations into monthly values. However, the tropical forest areas are masked out in this dataset [19] because of the difficulty of measuring soil moisture over dense vegetation with microwave remote sensing [23]. We noted a number of gridded soil moisture products with complete coverage of Africa, e.g., [24], but the soil moisture estimates are modeled from other variables, including precipitation. Thus, we opted for the microwave satellite soil moisture product, which is an independent set of measurements.

The GRACE system is made up of two satellites orbiting at identical orbital paths separated at a distance of ~220 km [25]. GRACE provides monthly measurements of the Earth's gravity field variations, as a function of local landmasses. After accounting for atmosphere and ocean effects [26], and taking into cognizance the relatively low contribution of vegetation biomass variations [27], the net mass variation can be attributed to the redistribution of TWS. Thus, TWS is an integral of surface water, soil moisture, and groundwater. In this study, we used an ensemble of average TWS from April 2002 to December 2015 computed from the GRACE data (release-5, level-2) independently pre-processed by three research centers (NASA Jet Propulsion Laboratory (JPL), University of Texas Center for Space Research (CSR), and the GeoForschungsZentrum (GFZ) Potsdam). However, GRACE's original signal is lost during pre-processing (e.g., filtering, de-stripping, and truncation). Hence, it is important that an appropriate scaling factor is applied to restore the signal loss prior to using water-related GRACE-TWS applications. For this reason, the scaling factor from the NCAR's Community Land Model 4.0 (CLM4.0, [28]) was applied to correct and restore the GRACE original signal. Besides, CLM4.0 takes into account the interaction between surface and groundwater in addition to accounting for human activities such as irrigation and river diversion [17,28,29]. Consequently, the TWS used in this study is measured in equivalent water height (EWH, in cm) at a spatial resolution of 1° and at a monthly time step. In order to match the January 2002 starting point of the soil moisture and precipitation datasets, the GRACE data was extended backward by replacing the three missing months (January, February, and March) with their long-term means. It is not expected that this data interpolation will have a significant impact on the result of this study. It is worth mentioning that the GRACE-TWS was derived from subtracting the monthly observations from a historical mean (2004–2009), the output of which is commonly referred to as terrestrial water storage anomalies (TWSA) in the GRACE user community.

The soil moisture, precipitation, and the GRACE datasets were resampled to co-register with the MODIS EVI data. Each pixel of the soil moisture (0.25°) and GRACE (1°) datasets were first disaggregated to 0.05° and then resampled by the nearest neighbor technique and then aligned to the EVI dataset. These two approaches ensure minimum loss of information in the downscaling process. However, the CHIRPS pixels were only resampled to match the MODIS EVI pixels using the nearest neighbor approach. The two-step (disaggregation and nearest neighbor interpolation) approach used here replicates the pixels without necessarily altering the original cell values.

We masked some pixels based on the following considerations. Any pixel missing more than 20% of data or where there are more than three consecutive months of missing data was masked out. Otherwise, we filled the missing values using a spline interpolation algorithm [30]. Furthermore, to focus the analysis on vegetated areas, pixels with time series having a mean EVI not exceeding 0.1 were masked out. EVI values below this threshold are indicative of bare soil or open water bodies [31].

A key focus of this study is to understand the relative importance of the hydrological measures as drivers of vegetation greenness, based on original and anomalies values. For this reason, monthly anomalies were computed by subtracting the climatology (long-term mean) of each calendar month from the corresponding monthly time series over the years covered in this study. The equation for this is given by

$$X_{anomaly}(i,j) = X(i,j) - \frac{1}{n}\sum_{j=1}^{n} X(i,j),\tag{1}$$

where X is the monthly observed variable, i is the month, j is the year, and n is the number of the years of the data. It should be noted that for the GRACE data, the computed monthly anomaly should not be confused with the original GRACE data in which the long-term mean has been subtracted and is referred to as TWSA. To avoid confusion, the original GRACE data will still be referred to as TWSA in this paper.

2.2. Trend Analysis

Trend analysis was performed on the time series of the monthly anomalies of each of the variables: EVI, precipitation, soil moisture, and TWSA over the period 2003 to 2015. The trend was estimated using the Mann–Kendell (MK) trend test [32,33]. The MK test evaluates a series for the presence and persistence of monotonic trend (increase/decrease) through pairwise comparison of observations. The test outputs Kendall's tau rank correlation coefficients (τ), which takes on values between -1 and 1 [33]. Positive, zero, or negative τ values, respectively, indicate an increasing, no trend, or a decreasing trend. The test is non-parametric and is widely used in remote sensing time series analysis, e.g., [12]. The MK test was used in this study because of its robustness to outliers and its capacity to handle short or noisy series compared to parametric tests such as ordinary least squares regression. Finally, only Kendall's τ values where the estimated trend was significantly different from zero ($p < 0.05$) were retained.

2.3. Modeling the Relationship between Vegetation Greenness Dynamics and Water Availability

In cognizance of the possible vegetation response to water availability exhibiting time lags, the relationship between vegetation greenness and the water availability was estimated with EVI as the response variable and 0–6 and 12 months lagged measures of the hydrological variables as predictors. Consequently, the time series of EVI for the period 2003 to 2015 was used, whereas the predictor variables were drawn from the time series of the hydrological variables covering the period from 2002 to 2015. This resulted in 156 observations for each pixel for the response and the 24 predictor variables (3 variables $\times$ 8 lags).

The relationship between EVI and the hydrological measures at the associated lags was estimated with random forest (RF). We chose to use RF because it is a distribution-free, non-parametric algorithm that can be used to model simple and complex relationships between response and predictor variables [7]. It has been widely used in many regression and classification problems [34,35]. In addition to being robust in the presence of irrelevant features, RF estimates a metric for the relative importance of the predictor variables to the prediction problem [7]. The variable importance (VIP) metric indicates the decrease in prediction accuracy on an out-of-bag sample (test sample) in which the values of a variable are randomly reshuffled as against the prediction accuracy on the same out-of-bag sample with no reshuffle. The higher the decrease in prediction accuracy, the more is the predictive power of the variable in the model. In this study, the RF model was estimated for each pixel with the number of trees set to 500, above which there is no significant improvement of model performance. The default values for the other parameters (e.g., mtry and node size) in the R statistical software implementation of RF were used.

Model performance assessment: The strength of the model was assessed by comparing the out-of-bag prediction of EVI against the observed values. The comparison was based on Lin's concordance correlation coefficient (LCCC) metric [36]. LCCC expresses the degree of agreement between the predicted values with the observed values, with respect to the 1:1 line [36].

3. Results

3.1. General Trends in EVI, Soil Moisture, TWSA, and Precipitation

Broadly speaking, there was a dominant positive trend in vegetation greenness anomalies in most parts of Africa (Figure 2) over the study period. Nonetheless, large clusters of a negative trend in greenness anomalies can be observed in Algeria, Tunisia, Libya, Niger, Nigeria, Ghana, Angola, and in countries in the eastern flank of Africa, extending from Eretria southwards to South Africa, and eastwards to Madagascar. Interestingly, the trajectories of vegetation greenness anomalies spatially coincided with trends in one or more of the hydrological variables in some locations. For example, the positive trend in greenness anomalies in southern Mali, in the region around the Sudan/South Sudanese border, and in parts of Angola and South Africa, is matched by an upward trend in at least two of the hydrological variables. However, a diverging direction of anomalies trends between vegetation and particularly soil moisture and TWSA, can be observed in Libya, Egypt, and also in Zambia. Finally, an important trend of ecological note is the pronounced clusters of a negative trend in precipitation and TWSA anomalies in parts of the Congo Basin.

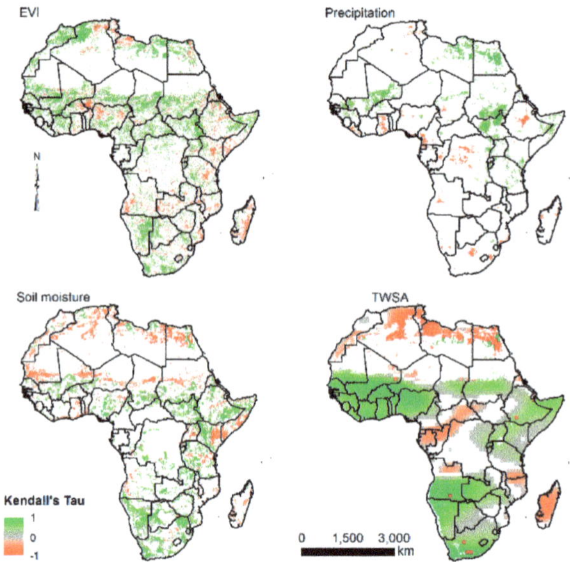

Figure 2. Trends in anomalies of EVI, precipitation, soil moisture, and TWSA over Africa from 2003 to 2015. The white areas are either no trend ($p > 0.05$), barren, no data, or water bodies.

3.2. Relationship between Vegetation Greenness Dynamics and Water Availability

Figure 3 shows the maps indicating the strength of the relationship between vegetation greenness and the hydrological variables and their lags as modeled with values of original observations (Figure 3A) and those of the monthly anomalies (Figure 3B). In the case of the original values, the relationship was very strong for most of Africa (LCCC > 0.75), although moderate relationships (LCCC = 0.5–0.75) were observed in Somalia, parts of Ethiopia and Kenya, southern Namibia, and Western South Africa, and in the northern margins of the Sahel. On the other hand, anomalies in vegetation greenness were generally less coupled to anomalies in water availability in many parts of Africa (LCCC < 0.25), notably in areas quite north of the Equator, predominantly across the savanna zones from west Senegal to the east coast of Eretria and Djibouti. However, in Namibia, Botswana, Kenya, Somalia and parts of Tanzania, Ethiopia, and South Africa, a moderate relationship between anomalies in vegetation greenness and water availability was evident (LCCC = 0.5–0.75).

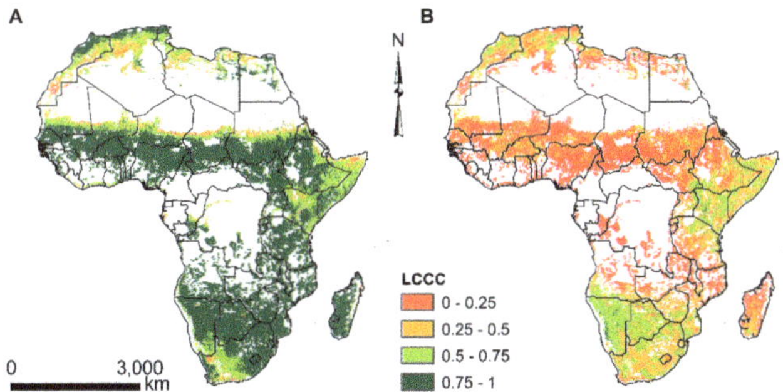

Figure 3. Spatial variability of the strength (LCCC) of vegetation greenness response to concurrent and lagged precipitation, soil moisture, and TWSA for the period 2003 to 2015 based on monthly time series of (**A**) original observations and (**B**) anomalies. The white areas are masked out either due to no data, barren land, or water bodies.

Figure 4 summarizes the relationship between vegetation greenness and the hydrological variables and their lags across the major land cover types. Again, the relationship was stronger for all land cover types in the case of original values (Figure 4A) in comparison to the anomalies (Figure 4B). Furthermore, relative to the other land cover types, the relationship was generally stronger in the woodlands and croplands with the original values, whereas vegetation greenness anomalies were more coupled with anomalies in water availability in the grasslands than in the other land cover types. Regardless of which data type (original or anomaly values) was used, vegetation greenness in the forest class was relatively less associated with water availability than in the other land cover types (Figure 4A,B).

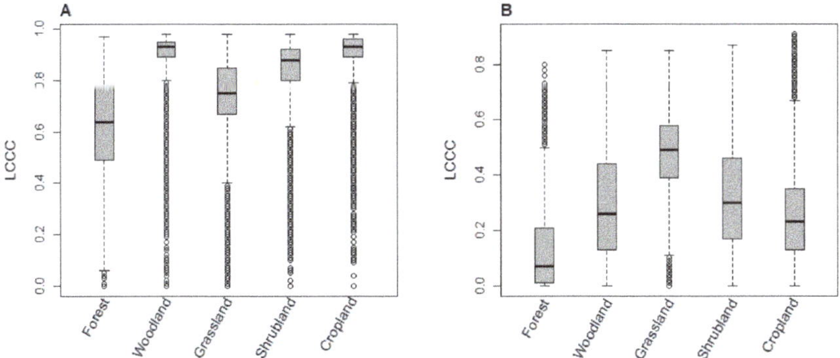

Figure 4. Relationship between vegetation greenness and concurrent and lagged precipitation, soil moisture, and TWSA across major land cover types for the period 2003 to 2015 based on monthly time series of (**A**) original observations and (**B**) anomalies.

3.3. Relative Importance of Soil Moisture, Precipitation, and TWSA as Drivers of Vegetation Greenness Dynamics

Figure 5 shows the two top-ranked predictor variables and their associated lags from the modeled relationship between vegetation greenness and water availability for the original (Figure 5A) and monthly anomalies (Figure 5B) data. Figure 6 further illustrates the proportion of pixels across the five top-ranked predictors. Based on the original observations, precipitation and soil moisture were the most important predictors of vegetation greenness. The proportion of pixels where soil moisture was ranked as the topmost VIP in predicting vegetation greenness was about 55% (Figure 6A). This encompassed parts of west southern Africa, East Africa, and the western Maghreb (Figure 5A). However, the proportion rapidly declined to about 30% in the fifth VIP. Precipitation, on the other hand, was the most important hydrological variable driving vegetation greenness in the eastern flank of southern Africa (Figure 5A). The proportion of pixels across Africa where precipitation was ranked among the top five VIP was fairly constant (45%, Figure 6A). In terms of vegetation lag response to the hydrological variables, and based on the first two important predictors, precipitation generally led vegetation greenness by 1 to 2 months (e.g., in eastern South Africa), whereas soil moisture was more concurrent with or led vegetation greenness by 1 month (e.g., in Namibia and Botswana).

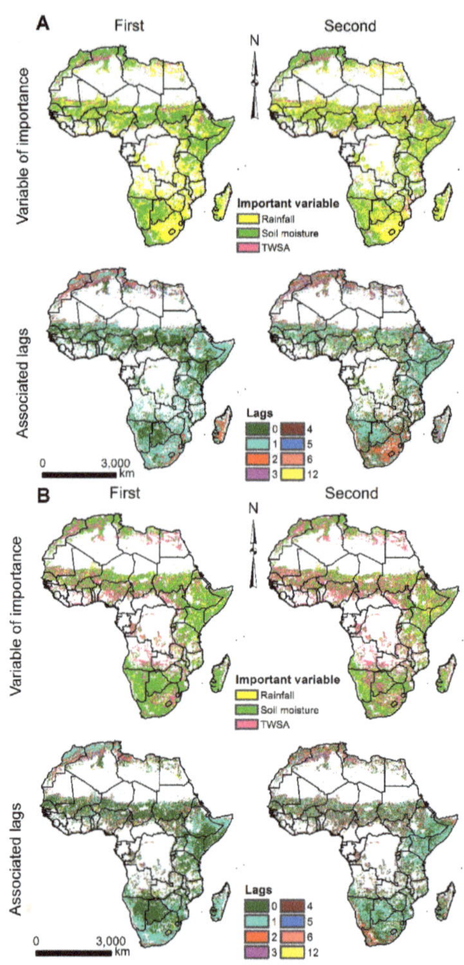

Figure 5. Spatial distribution of the top two ranked VIPs of the hydrological variables and their associated lags in (months) in modeling vegetation response to water availability in Africa based on monthly time series of (**A**) original observations and (**B**) anomalies. The white areas are masked out either due to no data, barren land, or water bodies.

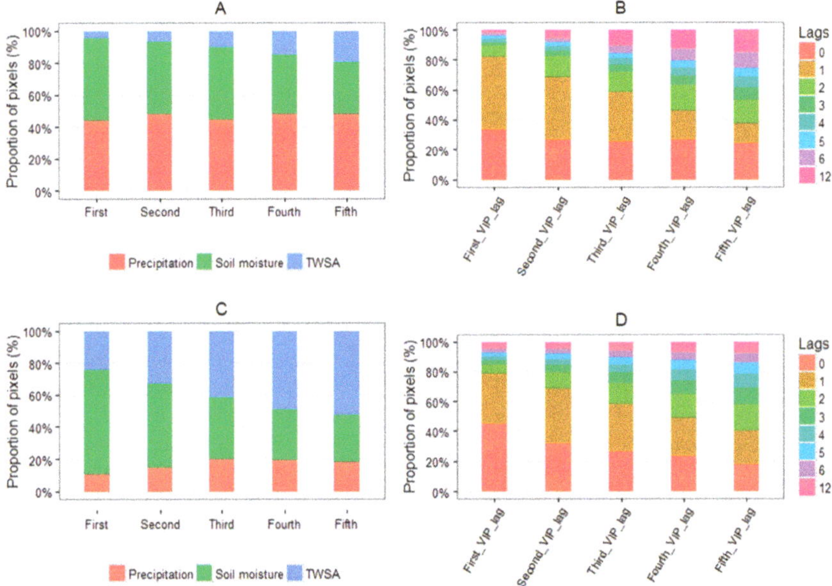

Figure 6. Summary of the first five ranked variable of importance of the hydrological variables and their lags in driving vegetation response to water availability across Africa based on (**A,B**) original and (**C,D**) anomalies monthly data.

Anomalies in soil moisture and TWSA were the dominant hydrological predictors of anomalies in vegetation greenness in most of Africa, with the role of precipitation greatly diminished (Figure 6C). Whereas soil moisture anomalies drive anomalies in vegetation greenness in virtually all parts of Africa, anomalies in TWSA were more coupled to vegetation dynamics anomalies in the humid savanna zone of Nigeria and the semi-arid region of Egypt (Figure 5B). Nonetheless, the proportion of pixels with dominant soil moisture control of vegetation greenness anomalies decreased considerably from the first (65%) to the fifth (30%) ranked VIP. This is in contrast to that of TWSA that increased from 25% to 50% (Figure 6D). Going by the first and second VIP results, vegetation anomalies lagged precipitation and TWSA anomalies by 0 to 1 month, though the response was much delayed in some parts of North Africa (Figure 5B).

4. Discussion

4.1. Trends in EVI and the Hydrological Variables

Some of the observed trends in vegetation greenness in this study are consistent with those reported by more recent regional and continental studies. For example, the positive trend in vegetation greenness observed in most parts of the Sahel has also been reported in recent studies by Leroux et al. [37], which examined vegetation changes in the Sahel between 2000 and 2015, and by Ugbaje et al. [12], which investigated the variability of vegetation productivity across Africa between 2000 and 2014. Additionally, the hotspot of declining vegetation greenness observed in southwestern Niger is consistent with findings from many studies (e.g., [37,38]). This hotspot has been dubbed "a Sahelian exception" because it stands in contrast to the dominant Sahelian greening [37]. Likewise, the negative trend in greenness over Zambia is in line with the reported decline in vegetation productivity by Ugbaje et al. [12] for the period between 2000 and 2014. Similarly, the clusters of a negative trend in vegetation greenness in parts of Somalia, Kenya, and Tanzania correspond to those reported by [39] over the period 2000 to 2010. This indicates that vegetation activity in this region has continued to decline even after 2010, up to 2015, as reported here.

Similar to the observed vegetation greenness trends, some of the trajectories of hydrological variables are in line with those of other studies. For instance, the positive trend in soil moisture and precipitation in southern Chad and South Sudan was also reported in a study by Huber et al. [9] which analyzed vegetation greenness dynamics in relation to water availability in the Sahel over the period of 1982 to 2007. Similarly, the general positive trend in soil moisture in southern Africa is in agreement with the improvement in soil water content for the period of 1993 to 2012, observed by Wei et al. [40]. Thus, our results indicate that this positive trend in soil moisture and precipitation persisted up to 2015. Also, the observed decline in precipitation and TWSA in parts of the Congo basin has also been reported by Zhou et al. [41] for the period covering 2000 to 2012.

However, the results here show a few notable areas of differences in trend directions from what is reported by other studies. For example, in Ugbaje et al. [12], vegetation productivity in Angola was largely stable, which is in contrast with a dominant positive trend in vegetation greenness observed in this study (Figure 2). Similarly, contrary to Zhou et al. [41] who observed a corresponding decrease in EVI following a decline in water availability in the Congo, here, EVI for most parts of the basin was either stable or showed a positive trend (Figure 2). This may be due to improvement in non-water related constraints like decreasing cloud cover and the concomitant increase in solar radiation [42]. Overall, the differences in the trajectories of vegetation dynamics between these studies and those in this study may be linked to the contrast in the time interval considered for the trend analyses.

4.2. Spatiotemporal Variation of Vegetation Greenness in Relation to Water Availability

Unlike many studies that assessed the relationship of vegetation greenness and water availability using one or more hydrological variables from remote sensing (e.g., [3,6,43]), this is the first study, to the best of our knowledge, that has analyzed vegetation greenness dynamics in relation to precipitation, soil moisture, and TWSA within a multivariate framework. The RF modeling results of water availability were generally strongly coupled to vegetation greenness when the cyclic seasonal variation in the analyzed time series is not removed. This is generally expected for Africa as the vegetation phenological cycle is primarily driven by the wet–dry season cycle. However, the moderate relationship observed in the semi-arid areas like in the Horn of Africa and the northern Sahel can be explained by their erratic precipitation regime.

On the flip side, when the monthly means were removed from the time series, the strong relationships between vegetation greenness and water availability was not as widespread across Africa. Most notably, the observed relationship was weak in the savanna belt stretching from the West coast of Africa to the east coast of Eretria and Djibouti. This result implies that the predominantly positive trend in greenness anomalies observed in the region may be driven by factors other than water availability. Such factors may include atmospheric fertilization, agricultural intensification including the use of improved and high yielding crop cultivars, afforestation, and a positive trend in the growing season length [44–46]. However, in eastern and southern Africa, south of Zambia, vegetation greenness anomaly dynamics was more linked to anomalies in water availability compared to the relationship observed in West Africa across to the east coast of Eretria and Djibouti. The relatively high proportion of grass cover in most parts of eastern and southern Africa (see Figure 1) may, in part, explain this difference in regional response to water availability. Because of their shallow rooting system, grasses are generally more sensitive to fluctuations in near-surface soil moisture availability than most woody plants [47], as can be seen from the relatively strong relationship obtained for the grassland cover type (Figure 4B). Nonetheless, the sensitivity of grassland linked to moisture anomalies is also a reflection of the increasing frequency and severity of water-related extreme events such as drought and flooding in eastern and southern Africa [48,49]. Therefore, the dynamics in water availability related to anomalies in one or more of the hydrological variables (soil moisture, TWSA, and rainfall) was to a large extent the principal driver of trends in vegetation greenness anomalies observed in eastern and southern Africa.

4.3. The Relative Roles of Precipitation, Soil Moisture and TWSA in Driving Vegetation Greenness

The dominance of soil moisture as the first ranked predictor of vegetation greenness dynamics in substantial parts of Africa from the two RF models (original or monthly anomalies data) (Figure 5) can be attributed to soil moisture being a better integral of the effects of topography and energy on plant available water than precipitation and TWSA. This also confirms soil moisture as a vital component of the hydrological cycle better linked with the vegetation phenological cycle [8]. However, there are locations where precipitation and/or TWSA were better indicators of plant available water than soil moisture. For example, with the original data, precipitation outranked soil moisture (and TSWA) as a predictor of vegetation greenness dynamics in south-eastern Africa which contrasts with south-western Africa where soil moisture exercised dominant control (Figure 5A). The better performance of precipitation in south-eastern Africa can be attributed to the strong influence of the adjacent Indian Ocean and ENSO events on the ecohydrological regime of the area [50].

Although precipitation is considered the primary climatic driver of vegetation phenology [8], here monthly anomalies of precipitation were not as important as anomalies in soil moisture and TWSA in explaining anomalies in vegetation canopy greening across Africa. This result is similar to the findings in other studies that reported a better correlation between satellite-observed soil moisture and NDVI anomalies (e.g., [4]) and between GRACE TWSA and NDVI anomalies (e.g., [6]). This result, therefore, is in line with the understanding that the amount of water stored in the soil rather than the amount of precipitation received determines the survival of plants to extreme events like drought.

Regardless of whether the original or anomalies time series was used for modeling, vegetation greenness generally responded quicker (0–1 months) to changes in soil moisture and TWSA than to precipitation (1–2 months). This is plausible especially if we consider precipitation as the source and soil moisture and TWSA as the conduit of plant-available water. This premise is also supported by the results of Chen et al. [4] and Yang et al. [6] which indicated a generally strong correlation between vegetation greenness anomalies and soil moisture and TWSA anomalies over Australia at 0 to 1 months of vegetation delayed response. On the global scale, Xie et al. [51] also observed a vegetation greenness delayed response of 0- to 1-month to the TWSA. Additionally, Yang et al. [6] and Gessner et al. [3] found NDVI to be strongly correlated with precipitation at time lags upwards of a month.

5. Conclusions and Outlook

Remote sensing observations of EVI, soil moisture, TWSA, and station-satellite blend precipitation products were used to assess the impact/influence of the hydrological controls of vegetation greenness dynamics over Africa for the period between 2003 and 2015. By using a multivariate approach with the distribution-free RF algorithm, the relationships between these hydrological variables and vegetation greenness, with and without monthly anomalies, were assessed. The advantage of this approach is that it allows for the modeling of complex interactions between vegetation greenness and the hydrological variables, including lag effects. This is a more robust way to assess the relative importance of the hydrological variables and their lag effects on vegetation response to water availability.

In most parts of Africa, the RF model with the seasonal component present in the time series (original data) generally performed better than the model driven by monthly anomalies of the variables. These results indicate that water availability is a better driver of vegetation phenology than anomalous vegetation greenness trends in most parts of Africa. This contrast in model performance is particularly striking across West Africa and eastward to Eretria and Djibouti. With regards to the relative importance of the three hydrological variables to vegetation greenness dynamics, soil moisture was the only variable that consistently performed with both time series types and was ranked as the first important variable in more than 50% of the pixels examined. Nonetheless, precipitation and TWSA had significant roles in controlling vegetation greenness with the original and monthly anomaly time series, respectively. Of particular note is the strong predictive power of precipitation in the eastern flank of south-eastern Africa, which is a demonstration of the strong influence of the adjacent Indian Ocean on the vegetation dynamics of the region. In terms of the response of vegetation greenness to changes in the hydrological measure, soil moisture, and TWSA were generally concurrent or led vegetation by 1 month, whereas precipitation led vegetation by 1–2 months. This demonstrates that soil moisture and TWSA are direct indicators of plant-available water than precipitation.

However, a key point to note is that soil moisture may have masked the strength of TWSA in predicting vegetation greenness response to water since TWSA is a measure of water stored from the surface to the boundary between the earth's crust and the mantle, which is inclusive of soil moisture. The soil moisture product used in this study represents measurements not deeper than 10 cm. This relationship between soil moisture and TWSA explains why the total area where soil moisture was important rapidly declined with a concomitant increase in the total area of TWSA influence across the variable of importance ranking (Figure 6). Thus, there is the need to further partition GRACE TWSA into the surface and groundwater components. However, because the current microwave sensors measure only soil moisture at the top few centimeters, TWSA supplements these measurements in areas where the root zone is deeper. Another possibility is to jointly assimilate the SMOS near-surface soil moisture observations and TWSA into a hydrological model to better approximate the root zone soil moisture, as demonstrated recently by Tian et al. [47], providing a better indicator of plant-available water and better vegetation response.

All in all, our results illustrate the usefulness of remote sensing soil moisture and TWSA as complementary data to precipitation in assessing and monitoring vegetation greenness dynamics, despite their relatively low spatial resolution (>0.25°). This is especially important in Africa, where there is a dearth of in situ precipitation and soil moisture observations. Furthermore, the relationships observed across the vegetation types in this study can be used in benchmarking coupled vegetation-climate models. The generally low correlation between vegetation greenness anomalies and the hydrological variables anomalies observed over most of Africa (Figure 3B) can be improved if additional hydroclimatic variables such as vapor pressure deficit and evapotranspiration, as well as root zone soil moisture estimates, are incorporated to the RF model.

Author Contributions: Conceptualization, S.U.U. and T.F.A.B.; methodology, S.U.U. and T.F.A.B.; formal analysis, S.U.U.; software, S.U.U.; writing—original draft preparation, S.U.U.; writing—review and editing, T.F.A.B.; supervision, T.F.A.B. All authors have read and agreed to the published version of the manuscript.

Funding: This research received no external funding.

Conflicts of Interest: The authors declare no conflict of interest.

References

1. Heimann, M.; Reichstein, M. Terrestrial ecosystem carbon dynamics and climate feedbacks. *Nature* **2008**, *451*, 289–292. [CrossRef] [PubMed]

2. Herrmann, S.M.; Anyamba, A.; Tucker, C.J. Recent trends in vegetation dynamics in the African Sahel and their relationship to climate. *Glob. Environ. Chang.* **2005**, *15*, 394–404. [CrossRef]

3. Gessner, U.; Naeimi, V.; Klein, I.; Kuenzer, C.; Klein, D.; Dech, S. The relationship between precipitation anomalies and satellite-derived vegetation activity in Central Asia. *Glob. Planet. Chang.* **2013**, *110*, 74–87. [CrossRef]

4. Chen, T.; McVicar, T.R.; Wang, G.; Chen, X.; De Jeu, R.A.M.; Liu, Y.Y.; Shen, H.; Zhang, F.; Dolman, A.J. Advantages of using microwave satellite soil moisture over gridded precipitation products and land surface model output in assessing regional vegetation water availability and growth dynamics for a lateral inflow receiving landscape. *Remote Sens.* **2016**, *8*, 428. [CrossRef]

5. Ndehedehe, C.E.; Ferreira, V.G.; Agutu, N.O. Hydrological controls on surface vegetation dynamics over West and Central Africa. *Ecol. Indic.* **2019**, *103*, 494–508. [CrossRef]

6. Yang, Y.; Long, D.; Guan, H.; Scanlon, B.R.; Simmons, C.T.; Jiang, L.; Xu, X. GRACE satellite observed hydrological controls on interannual and seasonal variability in surface greenness over mainland Australia. *J. Geophys. Res. Biogeosci.* **2014**, *119*, 2245–2260. [CrossRef]

7. Breiman, L. Random forests. *Mach. Learn.* **2001**, *45*, 5–32. [CrossRef]

8. Stampoulis, D.; Andreadis, K.M.; Granger, S.L.; Fisher, J.B.; Turk, F.J.; Behrangi, A.; Ines, A.V.; Das, N.N. Assessing hydro-ecological vulnerability using microwave radiometric measurements from WindSat. *Remote Sens. Environ.* **2016**, *184*, 58–72. [CrossRef]

9. Huber, S.; Fensholt, R.; Rasmussen, K. Water availability as the driver of vegetation dynamics in the African Sahel from 1982 to 2007. *Glob. Planet. Chang.* **2011**, *76*, 186–195. [CrossRef]

10. Campo-Bescós, M.; Muñoz-Carpena, R.; Southworth, J.; Zhu, L.; Waylen, P.; Bunting, E. Combined Spatial and Temporal Effects of Environmental Controls on Long-Term Monthly NDVI in the Southern Africa Savanna. *Remote Sens.* **2013**, *5*, 6513–6538. [CrossRef]

11. Farrar, T.J.; Nicholson, S.E.; Lare, A.R. The influence of soil type on the relationships between NDVI, rainfall, and soil moisture in semiarid Botswana. II. NDVI response to soil moisture. *Remote Sens. Environ.* **1994**, *133*, 121–133. [CrossRef]

12. Ugbaje, S.U.; Odeh, I.O.A.; Bishop, T.F.A.; Li, J. Assessing the spatio-temporal variability of vegetation productivity in Africa: Quantifying the relative roles of climate variability and human activities. *Int. J. Digit. Earth* **2017**, *10*, 879–900. [CrossRef]

13. Didan, K. MOD13C2 MODIS/Terra Vegetation Indices Monthly L3 Global 0.05Deg CMG V006. 2015. Available online: https://lpdaac.usgs.gov/products/mod13c2v006/ (accessed on 31 July 2017).

14. Funk, C.; Peterson, P.; Landsfeld, M.; Pedreros, D.; Verdin, J.; Shukla, S.; Husak, G.; Rowland, J.; Harrison, L.; Hoell, A.; et al. The climate hazards infrared precipitation with stations—A new environmental record for monitoring extremes. *Sci. Data* **2015**, *2*, 150066. [CrossRef] [PubMed]

15. EODC Product Specification Document (D1.2.1 Version 1.9). Available online: http://www.esa-soilmoisture-cci.org/sites/default/files/documents/ESA_CCI_SM_PSD_D1.2.1_v1.9.pdf (accessed on 1 July 2017).

16. Swenson, S.; Wahr, J. Post-processing removal of correlated errors in GRACE data. *Geophys. Res. Lett.* **2006**, *33*. [CrossRef]

17. Landerer, F.W.; Swenson, S.C. Accuracy of scaled GRACE terrestrial water storage estimates. *Water Resour. Res.* **2012**, *48*. [CrossRef]

18. Huete, A.; Didan, K.; Miura, T.; Rodriguez, E.P.; Gao, X.; Ferreira, L.G. MODIS_MOD13_NDVI_referenc. 2002, 83, 195–213. *Remote Sens. Environ.* **2002**, *83*, 195–213.

19. Liu, Y.Y.; Parinussa, R.M.; Dorigo, W.A.; De Jeu, R.A.M.; Wagner, W.; Van Dijk, A.I.J.M.; McCabe, M.F.; Evans, J.P. Developing an improved soil moisture dataset by blending passive and active microwave satellite-based retrievals. *Hydrol. Earth Syst. Sci.* **2011**, *15*, 425–436. [CrossRef]

20. Liu, Y.Y.; Dorigo, W.A.; Parinussa, R.M.; De Jeu, R.A.M.; Wagner, W.; McCabe, M.F.; Evans, J.P.; Van Dijk, A.I.J.M. Trend-preserving blending of passive and active microwave soil moisture retrievals. *Remote Sens. Environ.* **2012**, *123*, 280–297. [CrossRef]

21. Wagner, W.; Dorigo, W.; de Jeu, R.; Fernandez, D.; Benveniste, J.; Haas, E.; Ertl, M. Fusion of Active and Passive Microwave Observations To Create an Essential Climate Variable Data Record on Soil Moisture. *ISPRS Ann. Photogramm. Remote Sens. Spat. Inf. Sci.* **2012**, *7*, 315–321.

22. Dorigo, W.A.; Gruber, A.; De Jeu, R.A.M.; Wagner, W.; Stacke, T.; Loew, A.; Albergel, C.; Brocca, L.; Chung, D.; Parinussa, R.M.; et al. Evaluation of the ESA CCI soil moisture product using ground-based observations. *Remote Sens. Environ.* **2015**, *162*, 380–395. [CrossRef]

23. Wagner, W.; Blöschl, G.; Pampaloni, P.; Calvet, J.-C.; Bizzarri, B.; Wigneron, J.-P.; Kerr, Y. Operational readiness of microwave remote sensing of soil moisture for hydrologic applications. *Nord. Hydrol.* **2007**, *38*, 1. [CrossRef]

24. Fan, Y.; van den Dool, H. Climate Prediction Center global monthly soil moisture data set at 0.5° resolution for 1948 to present. *J. Geophys. Res. D Atmos.* **2004**, *109*, 1–8. [CrossRef]

25. Rodell, M.; Famiglietti, J.S.; Chen, J.; Seneviratne, S.I.; Viterbo, P.; Holl, S.; Wilson, C.R. Basin scale estimates of evapotranspiration using GRACE and other observations. *Geophys. Res. Lett.* **2004**, *31*. [CrossRef]

26. Tapley, B.D. GRACE Measurements of Mass Variability in the Earth System. *Science* **2004**, *305*, 503–505. [CrossRef]

27. Rodell, M.; Chao, B.F.; Au, A.Y.; Kimball, J.S.; McDonald, K.C. Global biomass variation and its geodynamic effects: 1982-98. *Earth Interact.* **2005**, *9*, 1–19. [CrossRef]

28. Gent, P.R.; Danabasoglu, G.; Donner, L.J.; Holland, M.M.; Hunke, E.C.; Jayne, S.R.; Lawrence, D.M.; Neale, R.B.; Rasch, P.J.; Vertenstein, M.; et al. The community climate system model version 4. *J. Clim.* **2011**, *24*, 4973–4991. [CrossRef]

29. Long, D.; Longuevergne, L.; Scanlon, B.R. Global analysis of approaches for deriving total water storage changes from GRACE satellites. *Water Resour. Res.* **2015**, *51*, 2574–2594. [CrossRef]

30. Moritz, S.; Bartz-Beielstein, T. imputeTS: Time Series Missing Value Imputation in R. *R Journal* **2017**, *9*, 207–218. [CrossRef]

31. Myneni, R.B.; Williams, D.L. On the relationship between FAPAR and NDVI. *Remote Sens. Environ.* **1994**, *49*, 200–211. [CrossRef]

32. Mann, H.B. Nonparametric tests against trend. *Econom. J. Econom. Soc.* **1945**, *13*, 245–259. [CrossRef]

33. Kendall, M.G. A new measure of rank correlation. *Biometrika* **1938**, *30*, 81–93. [CrossRef]

34. Hengl, T.; Heuvelink, G.B.M.; Kempen, B.; Leenaars, J.G.B.; Walsh, M.G.; Shepherd, K.D.; Sila, A.; MacMillan, R.A.; De Jesus, J.M.; Tamene, L.; et al. Mapping soil properties of Africa at 250 m resolution: Random forests significantly improve current predictions. *PLoS ONE* **2015**, *10*, e0125814. [CrossRef] [PubMed]

35. Gessner, U.; Machwitz, M.; Esch, T.; Tillack, A.; Naeimi, V.; Kuenzer, C.; Dech, S. Multi-sensor mapping of West African land cover using MODIS, ASAR and TanDEM-X/TerraSAR-X data. *Remote Sens. Environ.* **2015**, *164*, 282–297. [CrossRef]

36. Lin, L.I. A concordance correlation coefficient to evaluate reproducibility. *Biometrics* **1989**, *45*, 255–268. [CrossRef]

37. Leroux, L.; Bégué, A.; Lo Seen, D.; Jolivot, A.; Kayitakire, F. Driving forces of recent vegetation changes in the Sahel: Lessons learned from regional and local level analyses. *Remote Sens. Environ.* **2017**, *191*, 38–54. [CrossRef]

38. Dardel, C.; Kergoat, L.; Hiernaux, P.; Mougin, E.; Grippa, M.; Tucker, C.J.J. Re-greening Sahel: 30 years of remote sensing data and field observations (Mali, Niger). *Remote Sens. Environ.* **2014**, *140*, 350–364. [CrossRef]

39. Hoscilo, A.; Balzter, H.; Bartholomé, E.; Boschetti, M.; Brivio, P.A.; Brink, A.; Clerici, M.; Pekel, J.F. A conceptual model for assessing rainfall and vegetation trends in sub-Saharan Africa from satellite data. *Int. J. Climatol.* **2015**, *35*, 3582–3592. [CrossRef]

40. Wei, F.; Wang, S.; Fu, B.; Wang, L.; Liu, Y.Y.; Li, Y. African dryland ecosystem changes controlled by soil water. *Land Degrad. Dev.* **2019**, *30*, 1564–1573. [CrossRef]

41. Zhou, L.; Tian, Y.; Myneni, R.B.; Ciais, P.; Saatchi, S.; Liu, Y.Y.; Piao, S.; Chen, H.; Vermote, E.F.; Song, C.; et al. Widespread decline of Congo rainforest greenness in the past decade. *Nature* **2014**, *509*, 86–90. [CrossRef]

42. Nemani, R.R.; Keeling, C.D.; Hashimoto, H.; Jolly, W.M.; Piper, S.C.; Tucker, C.J.; Myneni, R.B.; Running, S.W. Climate-driven increases in global terrestrial net primary production from 1982 to 1999. *Science* **2003**, *300*, 1560–1563. [CrossRef]

43. Chen, T.; de Jeu, R.A.M.; Liu, Y.Y.; van der Werf, G.R.; Dolman, A.J. Using satellite based soil moisture to quantify the water driven variability in NDVI: A case study over mainland Australia. *Remote Sens. Environ.* **2014**, *140*, 330–338. [CrossRef]

44. Boschetti, M.; Nutini, F.; Brivio, P.A.; Bartholomé, E.; Stroppiana, D.; Hoscilo, A. Identification of environmental anomaly hot spots in West Africa from time series of NDVI and rainfall. *ISPRS J. Photogramm. Remote Sens.* **2013**, *78*, 26–40. [CrossRef]

45. Vrieling, A.; de Leeuw, J.; Said, M. Length of Growing Period over Africa: Variability and Trends from 30 Years of NDVI Time Series. *Remote Sens.* **2013**, *5*, 982–1000. [CrossRef]

46. Pan, S.; Dangal, S.R.S.; Tao, B.; Yang, J.; Tian, H. Recent patterns of terrestrial net primary production in Africa influenced by multiple environmental changes. *Ecosyst. Heal. Sustain.* **2015**, *1*, 1–15. [CrossRef]

47. Tian, S.; Renzullo, L.J.; Van Dijk, A.I.J.M.; Tregoning, P.; Walker, J.P. Global joint assimilation of GRACE and SMOS for improved estimation of root-zone soil moisture and vegetation response. *Hydrol. Earth Syst. Sci.* **2019**, *23*, 1067–1081. [CrossRef]

48. Hooli, L.J. Resilience of the poorest: Coping strategies and indigenous knowledge of living with the floods in Northern Namibia. *Reg. Environ. Chang.* **2016**, *16*, 695–707. [CrossRef]

49. Nicholson, S.E. A detailed look at the recent drought situation in the Greater Horn of Africa. *J. Arid Environ.* **2014**, *103*, 71–79. [CrossRef]

50. Nicholson, S.E.; Kim, J. The relationship of the el nino oscillation to african rainfall. *Int. J. Climatol.* **1997**, *17*, 117–135. [CrossRef]
51. Xie, X.; He, B.; Guo, L.; Miao, C.; Zhang, Y. Detecting hotspots of interactions between vegetation greenness and terrestrial water storage using satellite observations. *Remote Sens. Environ.* **2019**, *231*, 111259. [CrossRef]

 *land*

Article

An Assessment of Multiple Drivers Determining Woody Species Composition and Structure: A Case Study from the Kalahari, Botswana

Thoralf Meyer [1], Paul Holloway [2,3,*], Thomas B. Christiansen [1], Jennifer A. Miller [1], Paolo D'Odorico [4] and Gregory S. Okin [5]

1   Department of Geography and then Environment, University of Texas at Austin, Austin, TX 78712, USA
2   Department of Geography, University College Cork, Cork T12K8AF, Ireland
3   Environmental Research Institute, University College Cork, Cork T23XE10, Ireland
4   Department of Environmental Science, Policy, & Management, University of California Berkeley, Berkeley, CA 94720, USA
5   Department of Geography, University of California Los Angeles, CA 90095, USA
*   Correspondence: paul.holloway@ucc.ie; Tel.: +353-490-2835

Received: 29 June 2019; Accepted: 1 August 2019; Published: 5 August 2019

Abstract: Savannas are extremely important socio-economic landscapes, with pastoralist societies relying on these ecosystems to sustain their livelihoods and economy. Globally, there is an increase of woody vegetation in these ecosystems, degrading the potential of these multi-functional landscapes to sustain societies and wildlife. Several mechanisms have been invoked to explain the processes responsible for woody vegetation composition; however, these are often investigated separately at scales not best suited to land-managers, thereby impeding the evaluation of their relative importance. We ran six transects at 15 sites along the Kalahari transect, collecting data on species identity, diversity, and abundance. We used Poisson and Tobit regression models to investigate the relationship among woody vegetation, precipitation, grazing, borehole density, and fire. We identified 44 species across 78 transects, with the highest species richness and abundance occurring at Kuke (middle of the rainfall gradient). Precipitation was the most important environmental variable across all species and various morphological groups, while increased borehole density and livestock resulted in lower bipinnate species abundance, contradicting the consensus that these managed features increase the presence of such species. Rotating cattle between boreholes subsequently reduces the impact of trampling and grazing on the soil and maintains and/or reduces woody vegetation abundance.

Keywords: conservation; fire; grazing; savanna; woody vegetation

---

1. Introduction

Unprecedented changes in climate, urbanization, and economic development are increasing the pressures that societies are enforcing on ecosystems [1]. Developing sustainable ecosystem services is subsequently a priority for conservation management, with savanna ecosystems a landscape of primary concern. Savannas are mixed plant communities comprised of grasses and woody vegetation that cover approximately a quarter of the Earth's land surface, including roughly half of the African continent [2]. Savannas are an extremely important socio-economic landscape in Africa, with over 80% of savanna land used to raise livestock [3], underpinning the economic stability of many countries [4,5]. The dynamic nature of savannas means they are susceptible to changes, particularly shifts in plant community composition associated with an increase in woody vegetation [6,7]. A particularly concerning aspect of this increased density of woody vegetation is the reduction of grasses and herbs by encroaching woody species. These negative impacts are occurring at an increasingly frequent rate worldwide [8–10], which is a major threat to the ecosystem stewardship of these economically important landscapes.

The transition of savanna ecosystems to open shrubland across Botswana, and in particular the western part of the Kalahari, presents a considerable threat to the conservation of the economically important ranching industry. In order to develop adaptive management strategies, the underlying environmental drivers of woody vegetation species need to be better understood. By understanding the environmental drivers responsible for the diversity and abundance of woody vegetation, we can develop predictive models to identify 'high-risk' areas, and provide managers, farmers, and governments with decision support across savanna landscapes. Previous research addressing the ecological processes responsible for the observed vegetation patterns have often found conflicting results regarding the importance and significance of these environmental drivers [11–15], thus limiting the use of this knowledge as the basis for decision-making at a landscape scale. These differences will be discussed below in the context of savanna ecosystems.

1.1. Precipitation

Rainfall affects water availability, and this factor has been described as the most important determinant describing woody vegetation communities, particularly as it limits the amount of primary productivity within an area [16–20]. For example, in a continental study of African savannas, Sankaran et al. [16] identified that woody cover increased linearly with mean annual precipitation (MAP) above 150 mm until maximum woody cover was reached at 650 mm. Similarly, in a pot experiment studying the growth of *Acacia* (new *Senegalia* and *Vachellia* classifications) species, Kraaij and Ward [19] found that rainfall frequency was the most important factor affecting both germination and survival of seedlings. Joubert et al. [21] also found that at least two successive seasons of favorable rainfall was required for seed recruitment in *Senegalia mellifera*. While precipitation intensities [22], season lengths [23], and interactions with other factors (e.g., grazing [24]) all influence woody vegetation cover, the consensus is that MAP is the primary factor contributing to woody vegetation cover [10,16–20].

1.2. Grazing

The influence of grazing pressure as a driver for increased woody vegetation cover is a long established theory. Walter's [25] two-layered hypothesis proposes that in savannas, grasses dominate the top-most soil layers, while tree roots dominate lower layers. When grazing removes the grass cover, tree roots begin to dominate the upper layers and prevent the grasses from reestablishing. Studies have proven inconclusive for the two-layer hypothesis, finding evidence both in support [26–28] and in opposition [29–31]; however, while this theory is still accepted, the current consensus is that this hypothesis is too simplistic to represent the complex dynamic savanna processes [17].

1.3. Trampling

Another explanation for the increased abundance of woody vegetation is the effect of trampling. Trampling from the high frequency and density of pastoral farming causes significant declines in cyanobacterial soil crust [32,33]. Savannas are characterized by low soil nutrient content [34–36], although many areas have biological soil crusts that increase soil surface stability, thereby reducing nutrient loss by erosion and atmospheric nitrogen fixation [33]. Studies have found that the soil crust is greatly influenced by this pastoral trampling within 2 to 8 km of boreholes [37], and that *Acacia* (new *Senegalia* and *Vachellia* classifications) species are often found in higher abundances within areas closer to boreholes, due to their low palatability and the positive species-specific association between canopy and soil crust development [38]. Boreholes are narrow shafts drilled into the ground in order to extract water and are the primary source of water for livestock farmers in southern Africa. Furthermore, cattle rarely stray more than 13–18 km from these water sources in Africa [39], meaning areas closer to boreholes may have increased woody vegetation cover.

1.4. Fire

Fire is a factor that restricts woody vegetation diversity and abundance, preventing the formation of canopies [40–42] as well as removing seedlings and subsequently preventing the establishment of new trees [43]. Furthermore, for certain species fire can also kill the larger trees [44,45]. Seymour and Huyser [45] found that infrequent fires were enough to kill established *Vachellia erioloba* trees, which are an important keystone species in the region, meaning an increase in fire frequency could have implications on biodiversity. In unmanaged areas, the build-up of large quantities of grass biomass in the understory results in high-intensity fires that are capable of destroying juvenile trees [46]. For example, Sankaran et al. [11] studied the effect of fire return intervals on the percentage of woody cover in African savannas and found that a shorter return interval reduced established woody cover, which kept the community in a juvenile state by 'top-killing' seedlings. In managed landscapes, fires are not as frequent or intense enough to have a discernible impact on mature trees [40], and a common feature of savannas is the reduction of fires due to mitigation strategies [47]. However, Joubert et al. [48] note that fire is crucial to disrupt transition from grassy savanna to thicket, and that managers who prevent fires at this stage are likely to experience bush thickening in the future.

1.5. Research Gap and Questions

Variation in species characteristics is fundamental to understanding biogeographic patterns [49]. One reason for the possible lack of conclusive evidence explaining the main drivers of different woody vegetation patterns in previous research is the variation in how vegetation has been measured (e.g., single species, multiple species, richness, percent woody cover), as well as the differences in spatial scales of the previous studies (ranging from garden experiments to coarse continental extents). Assessing diversity as total species richness does not always adequately characterize the way in which species differ from each other, and it is these differences in traits, which often indicate that species respond in different ways to changes in the environment [50,51]. Alternatively, studying only one species in isolation could lead to species-specific results that are not generalizable to the larger system or to other species. Several mechanisms (outlined above) have been invoked to explain the processes responsible for woody vegetation composition; however, these are often investigated separately at scales not best suited to land-managers, thereby impeding the evaluation of their relative importance.

Subsequently, this study focuses on the vegetation composition of the Botswana Kalahari, with the aim to investigate the relative influence of the environmental drivers of woody vegetation at a regional scale. By classifying species into morphological groups based on shared physiological traits, the drivers of woody vegetation richness and abundance can be interpreted more meaningfully at a regional scale that is more appropriate for landscape management decisions. This study will explore three main questions: (1) what is the woody vegetation composition of the Kalahari in western Botswana? (2) What are the environmental drivers of woody species richness? and (3) what are the environmental drivers of woody species abundance?

2. Materials and Methods

2.1. Study Area

We conducted our research in western Botswana between 2009 and 2011 (Figure 1). We created a 950 km transect following the observed rainfall gradient along the western part of the Kalahari. This transect ran from Shakawe in the northwest of the country to Bokspits in the southwest of the country. Rainfall along the transect decreases from the north to south, ranging from a MAP of 550 mm to 350 mm [52]. Along this transect, we identified 15 regions (Figure 1) where we conducted multiple vegetation surveys. We selected regions on their accessibility and a minimum distance of 75 km to the previous region.

Figure 1. Location of the 15 regions along the Kalahari Transect where fieldwork was undertaken.

2.2. Data Collection

Vegetation was surveyed using the line interception transect (LIT) method. Within each region, we fixed six 100 m transects radially from a center point. For the dry season, the direction of the first transect was determined by a random number (between 0 and 360), and the further two transects were offset by 120 degrees. Transects of the wet season were spaced exactly between dry season transects, resulting in an offset of 60 degrees from the very first transect laid. Transects were placed 200 m from the center point to avoid over-sampling a small area. See Krebs [53] for a further description of the LIT methodology. We recorded all woody vegetation that was taller than 25 cm following the nomenclature provided by Palgrave [54], whereby average height, distance covered over the transect line, and distance and direction of the stem(s) were documented. Species richness and abundance were recorded at all transects, and species identity were recorded at all sites, with the exception of the wet season transects at Sites 1, 3, 4, and 5 due to uncertain species identification resulting from missing leaves. The results of the vegetation survey meant we had data from 78 transects for use in the statistical analysis.

Species were categorized into five morphological groups based on the classification guidelines outlined by Meyer et al. [55]. Morphological group I consisted of species characterized by bipinnate leaf structures and growth form ranging from multi-stemmed shrub like appearance to single-stemmed trees. Morphological group II included broad leaf species forming dense canopy structures where the majority of the growth form is either multi-stemmed (generally less than five stems) or single-stemmed. Morphological group III contained multi-stemmed broad leaf shrubs with closed canopies, seldom exceeding 2 m in height. In contrast, morphological group IV contained shrub species characterized by open canopies. Morphological Group V included relatively short shrub species (<1.5 m) with small, open canopies (<0.5 m in diameter). We also obtained data on precipitation, fire frequency, cattle density, and borehole locations that represent the possible drivers of diversity and abundance of woody vegetation (Table 1).

Table 1. Description of the environmental drivers used to explore the diversity and abundance of woody vegetation in western Botswana.

Variable	Description	Source
Mean Annual Precipitation	We derived mean annual precipitation (MAP) from the isopleth vector data representing rainfall conditions across the Kalahari.	[56]
Fire Frequency	We derived fire frequency using the MODIS direct broadcast burned area product (MCD64A1) as described in Giglio et al. [57]. Fire frequency product and generation outlined in Appendix A.	[58]
Grazing	We identified density of cattle using the latest available Department of Wildlife and National Parks aerial counts of wildlife. This survey was conducted during the dry season of 2005.	[59]
Borehole Density	We counted the number of boreholes within an eight-kilometer (based on Dougill et al. [37]) radius.	[60]

2.3. Data Analysis

We performed regression analysis in order to explore the environmental drivers of woody species richness and species abundances. Environmental variables were checked for multicollinearity using variance inflation factor, then standardized using z-scores in order to compare their relative influence on the ecological indicators. We performed all regression analyses using R 3.3.0. [61]. We selected regression analyses based on a preliminary evaluation of the data and their error distribution. Histogram exploration identified a mixture of Poisson and censored Gaussian distributions. We subsequently used a combination of generalized linear models with Poisson error distributions and Tobit regression models to analyze our data. For data that had a Poisson distribution, a Generalized Linear Model procedure with a Poisson error distribution and a log link function was used:

$$\log(y) = \beta_0 + \beta_1 X_1 + \ldots + \beta_n X_n \tag{1}$$

where y is the abundances, X_n is the nth predictor, and β_n is the Poisson regression coefficient.

A censored Gaussian distribution represents a dataset that has a normal error distribution, but has some limit, either from below or above. Ecological data is often collected with a large proportion of the observations just above zero, while data cannot extend below zero or above certain thresholds (e.g., percentage cover). Tobit regression overcomes this bias and has been shown to perform better than ordinary least squares (OLS) (e.g., [62]) and is widely used in criminology (e.g., [63]) and land use change research (e.g., [64]). Species richness of woody vegetation is censored at zero (i.e., there cannot be a species richness of −1), and so any parameter estimates obtained by conventional OLS would be biased. Developed by Tobin [65], the Tobit regression model fits a set of parameters to where the dependent variable is left-censored at zero:

$$y_i^* = x_i \beta + \varepsilon_i \tag{2a}$$

$$y_i = \begin{cases} 0 \text{ if } y_i^* \leq 0 \\ y_i^* \text{ if } y_i^* > 0 \end{cases} \tag{2b}$$

where the subscript $i = 1, 2, 3 \ldots n$, indicates the observation, y_i^* is an unobservable variable, x_i is a vector of explanatory variables, β is a vector of unknown parameters, and ε_i is the error term. To estimate the censored regression models, we used the censReg [66] and MaxLik [67] packages. Final models were selected based on Akaike Information Criterion (AIC) using both forwards and backwards stepwise selection of models. We investigated third and fourth order interactions, but these did not improve the final models. Therefore, our final models only include main effects and second order interactions.

3. Results

3.1. Woody Vegetation Surveys

We identified 44 woody plant species across the 78 transects where taxonomic information was recorded. We recorded the highest diversity at Kuke (site 7), with 21 species found in all six transects in the region (Figure 2a), and 13 species found along transect four during the wet season. In general, both richness and abundance decreased as data collection moved southwards which follows the precipitation gradient, although Kuke is the notable exception. We recorded the lowest total abundances for all six transects at NG5 (Site 6) and Bokspits (Site 15) and the highest abundances at Quangwa (Site 4) and Kuke (Site 7) (Figure 2b). Supplementary Information provides the data which includes geographic location (WGS 1984), a list of species recorded, and their morphological classification. We recorded eight species in Morphological Group I (bipinnate leaf structure), fourteen species in Morphological Group II (tall dense canopies), fifteen species in Morphological Group III (small dense canopy species), five species in Morphological Group IV (tall open canopies), and two species in Morphological Group V (small open canopies). Due to the low number of species recorded in Morphological Group V, these were withheld from the statistical analysis to prevent any generalization or over-fitting of the models.

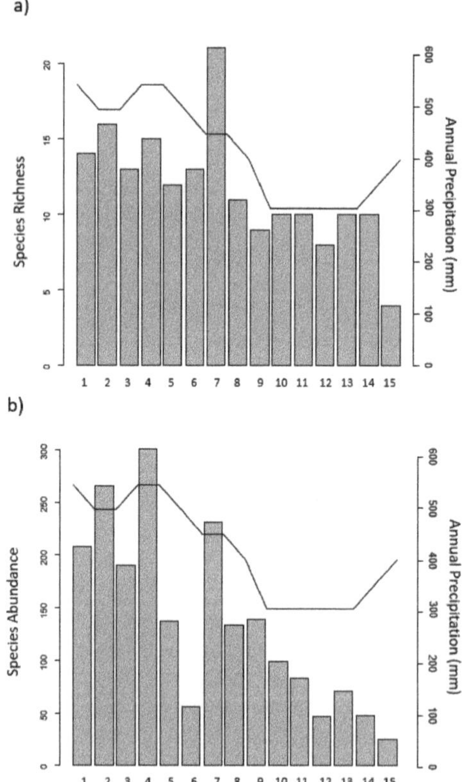

Figure 2. Total (**a**) woody vegetation species richness and (**b**) woody vegetation species abundance summed across the six transects at each of the 15 zones across the Kalahari Transect, plotted against annual precipitation (mm). Sites listed (1) Shakawe, (2) Tsodilo, (3) Gumare, (4) Quangwa, (5) Drotsky's Caves, (6) Ng 5, (7) Kuke, (8) Ghanzi, (9) Ghanzi South, (10) Bere, (11) Tshane, (12) Tshane South, (13) Mabuasehube, (14) Tsabong, and (15) Bokspits.

3.2. Regression Analysis

Our results indicated a number of important drivers of woody vegetation species richness and abundance (Table 2). Precipitation was the most important environmental variable when we considered all species together for both richness and abundance with borehole density and fire included in the final models. When species were deconstructed into morphological groups, we observed a variety of significant environmental drivers and interactions between these variables (** significant at $\alpha < 0.01$, * significant at $\alpha < 0.05$). The relative importance of each environmental driver often changed when we compared the regression models for richness and abundance of the same morphological group, indicating that the processes that determine diversity are different from those determining abundance. Boreholes were the most important driver for morphological groups II abundance and III richness, while livestock was the most important driver for morphological group IV abundance.

Table 2. Regression output for species richness (SR) and abundance (AB) for the four Morphological Groups (MG). ** significant at $\alpha < 0.01$, * significant at $\alpha < 0.05$. Tobit (T) and Poisson (P) regression analysis undertaken based on distribution of data.

	Total SR	Total AB	MG I SR	MG I AB	MG II SR	MG II AB	MG III SR	MG III AB	MG IV SR	MG IV AB
Regression	T	T	T	P	T	P	T	P	P	P
AIC	319.72	597.98	218.54	398.95	223.60	588.33	226.76	450.45	126.61	488.37
Intercept	5.66 **	20.05 **	0.92 **	−1.81	1.18 **	−0.05	1.00	0.34	−40.69	−4.67
PPT	1.32 **	11.07 **	−0.40 **	1.71 **	3.51 **	2.36 **	0.92 **	−0.01	−27.03	−4.32
Boreholes	−0.57 **	−1.85 **	−0.31	4.06 **	−1.65 **	−3.24 **	−2.78 **	−2.51 **	51.60	−0.04
Livestock				−12.97	1.14 **	2.04 **	1.62 **	1.87 **	−132.0	−10.02
Fire		−3.16 **	−0.20	−9.96	−0.60 **	−0.61 **		0.39	−14.52	0.13
*PPT * Boreholes*	0.56 **			5.80 **				−2.83 **	56.83	
*PPT * Livestock*				−2.59 **	4.17 **	2.37 **	0.79	1.68 **	−109.8	−8.30
*PPT * Fire*				−0.46 **		0.28		0.29 **	−2.44	
*Boreholes * Livestock*				−5.05 **		1.86 **	1.31	2.82 **		
*Boreholes * Fire*			−0.76					1.19	−27.79	
*Livestock * Fire*				−18.08						
logSigma	0.57 **	2.28 **	−0.01		0.27 **		0.07			

Precipitation and borehole density were included in all final models for every morphological group, while livestock was not important when all species were considered together, but included for all morphological groups (both richness and abundance) with the exception of morphological group I richness. Similarly, fire frequency was included for most morphological groups, with the exception of morphological group III and total species richness. Several two-way interactions were returned across the different models, and these were often significant. Morphological groups I and III abundance had the most interactions among all variables, suggesting these species have a complex and dynamic relationship with the environment.

Precipitation was a significant variable in all final regression models for species richness and abundance for all but three morphological groups, and it was the most important variable for total species richness and abundance, and morphological group I richness (Table 2). Precipitation had a positive relationship with species richness for morphological groups II and III, and abundance for morphological group II. This relationship was expected since these groups are characterized by dense canopy broad leaf species resulting in higher Leaf Area Index (LAI) and hence higher water requirements [68]. A negative relationship for morphological group I (bipinnate species) richness and rainfall was identified. This could be due to the fact that in xeric environments such species outcompete the majority of broad-leaved vegetation due to their general morphological characteristics and ecological traits (such as long root traps [69]), meaning the diversity of these species increases in arid areas where other water-dependent species simply cannot survive.

Livestock density was not included in the final model as selected by AIC when all species were considered together, but it had a positive relationship with morphological groups II and III (Table 2). Small dense canopy species such as *Grewia spp, Rhus tenuivirus* and *Ziziphus mucronata* notably have relatively low palatability [70]. Thus, if these species were already established when grazing increased in the area, they would not be affected by livestock. It was also the most important variable in determining abundance of morphological group IV (tall open canopy), reporting a negative relationship. Borehole density also had the most influence in determining both abundance of morphological group II and richness of morphological group III, forming a negative relationship with both ecological indicators. Boreholes also had a negative relationship with all response variables with the exception of morphological group I abundance. However, the interaction between boreholes and livestock density was significant for morphological group I abundance, indicating a negative relationship. This interaction was also significant for morphological groups II and III (although positive). These findings contradict previous research, and indicate that broad leaf species thrive in locations where there are more cattle and boreholes, while bipinnate species decrease. Fire had a negative influence on both richness and abundance at a regional scale (Table 2). Fire was generally negatively correlated to the overall abundance of woody species, but had a positive relationship with abundance of morphological groups III and IV, albeit not significant.

4. Discussion

Following the global trend in the conversion of savanna landscapes to woodier landscapes [7,27], the aim of this research was to investigate the variables responsible for woody vegetation composition in the western Kalahari, in particular those that cause high diversity and abundance of these species. We identified a variety of environmental drivers that are responsible for high diversity and abundance of woody vegetation, most notably precipitation, borehole density, grazing, and fire.

Our results generally agree with the observation that the rainfall gradient of the Kalahari is associated with an increase in woody vegetation [16–20]. Interestingly, the highest species richness was recorded at Kuke (Figure 1—Site 7), where the annual precipitation is 450 mm (in the middle of the rainfall gradient). The substantially higher species richness at Kuke can be explained by the site being located in an area buffering the Ghanzi farm-block to the south and the wildlife areas to the north. Both livestock and wildlife numbers are low here, and furthermore, fires have not occurred in this area due to both fire prevention strategies and the existence of the veterinary cordon fences acting as fire breaks. Therefore, our results indicate that while rainfall has a strong influence on woody vegetation, other factors also contribute significantly.

Our findings corroborate the positive association of bipinnate abundance (morphological group I) in areas close to boreholes [38], as well as an overall reduction in woody vegetation cover [71]. The negative relationship with small dense species is intuitive, as trampling loosens the soil and prevents these species from rooting. However, when grazing is high, the significant negative interaction between borehole density and grazing with bipinnate abundance contradicts the existing theories behind woody vegetation patterns. This relationship is a result of the fact that a higher number of boreholes and cattle represent more managed commercial ranches where cattle are routinely rotated between fields, and the regular use of multiple boreholes by the livestock negates the impact of trampling on the soil. This subsequently reduces the rate of bush encroachment by the unpalatable and thorny bipinnate species, and a positive relationship with other morphological groups is observed.

The negative relationship between fire frequency and woody vegetation corroborates observations from other dryland ecosystems [9,41] and supports a mechanistic understanding of the effect of fires in mixed tree-grass plant communities [40,72–74]. These findings support the observations at Kuke, that absence of fire does increase vegetation diversity and abundance (particularly for smaller species), and that the removal of fire from a landscape could increase bush thickening [49]. However, when fire was included in the models, it was seldom the most important variable (Table 2), with the exception of a positive interaction between livestock and fire when modelling morphological group I abundance (albeit not significant). While diversity and abundance did decrease, the lesser impact compared to the other environmental variables suggests that frequent fires may not have such severe implications on the ecosystem's biodiversity as proposed [45]. However, the MODIS MCD64A1 product used in this study ([57]; Appendix A) does not account for fire intensity which could still negatively impact the landscape.

The deconstruction of species into morphological groups that are internally homogenous provided an opportunity for an improved understanding of the processes that underlie the patterns [50]. Despite this, in savanna ecosystems, research has focused on individual species (e.g., [21,24,45]) where findings are generally not always scalable to the wider ecosystem as species do exhibit idiosyncratic responses to the environment [75]. Subsequently, we feel that our analysis has related the importance of environmental drivers on the structure and physiological properties of the species, while it is not so specific that we cannot generalize processes to a scale that is useful for land managers.

It should also be noted that other factors may influence woody vegetation patterns. Topographic heterogeneity [76], atmospheric carbon [46], and harvesting [77] have all been found to influence woody vegetation communities. These factors were excluded due to the topographically homogenous landscape under study, and the fact that regional data on carbon and harvesting are difficult to obtain; however, future research should continue to explore the impact of these factors. We also investigated time since last fire as a variable in the regression analysis; however, fire frequency was found to have more influence on woody vegetation patterns and was subsequently the only fire variable retained in the final models to prevent any issues of multicollinearity. Similarly, we measured grazing as density of cattle recorded from aerial surveys, although grazing could be represented using intensity (e.g., quantification of herbaceous tissue removal or an assessment of high, medium, or low). However, available data on such features was not available to this study. Recently, the statistical effects of spatial autocorrelation have been noted [78] and methods to incorporate and explore this into regression models have become more common [79–81]. However, we made the decision not to incorporate spatial autocorrelation in our analysis so that discussion could focus specifically on the environmental factors across the transect.

We used a combination of generalized linear models with Poisson error distributions and Tobit regression models to analyze our data. Biodiversity indicators such as species richness and abundance often exhibit distributions that are unsuitable for a number of statistical techniques. The literature surrounding the use of statistical analyses that do not account for lower limits to explore ecological questions is perhaps part of the reason we still have ambiguity surrounding the drivers of woody vegetation in savanna ecosystems. While our results corroborate the existence of well-established biodiversity-environment relationships (e.g., positive relationship with MAP), we also identified several novel biodiversity-environment relationships from the Tobit models (e.g., positive relationships with livestock). Subsequently, research should continue to explore more suitable statistical methodologies with which to analyze ecological data so that any management strategies implemented from findings are better informed.

5. Conclusions

Savannas are an extremely important socio-economic landscape in Africa. These landscapes are inherently multi-functional, balancing the needs of pastoral societies with conservation of these dynamic ecosystems. Global trends of savanna to shrubland conversion [6,7] will have important ecological and economic consequences. Here we investigated the impact of regional scale environmental drivers (a scale that is more relevant to governments and land managers across Africa and beyond) on woody vegetation diversity and abundance. Data on over 44 species was collected over a two-year period at fifteen sites along the Kalahari transect. At each site, six 100 m transects recorded diversity ranging from one species to thirteen species, and abundance ranging from two individuals to 62 individuals. A mixture of Poisson and Tobit regression models identified that rainfall was the most important environmental variable when all species were considered equally, corroborating previous research conducted at continental [11,16] and garden [17,19] scales. Interestingly, bipinnate species abundance decreased with increasing boreholes and livestock. These results contradict the consensus that borehole density and grazing increase the presence of such species, and suggest that by rotating cattle between boreholes, the impact of trampling and grazing on the soil is reduced and savanna landscapes are maintained. The deconstruction of species into different morphological groups provided better insights into the differences in the ways woody vegetation responds to environmental factors, and this deconstruction could aid in reconciling the divergent hypotheses surrounding woody vegetation patterns in savanna ecosystems, as all variables had a significant relationship with richness and abundance across all morphological groups. The results of this research should support land managers, governments and researchers working in transitional savanna landscapes worldwide.

Supplementary Materials: The following are available online at http://www.mdpi.com/2073-445X/8/8/122/s1, Data.

Author Contributions: Conceptualization, T.M., P.H., J.A.M., T.B.C., P.D.O., G.S.O.; Data Collection, T.M., P.D.O., G.S.O.; Data Analysis, T.M., P.H., J.A.M., T.B.C.; writing—original draft preparation, T.M., P.H., J.A.M., T.B.C., P.D.O., G.S.O.; writing—review and editing, T.M., P.H., J.A.M.

Funding: This research was funded by National Science Foundation, grant number DEB-0717448 supported T.M., P.D.O., and G.S.O., and grant number 0962198 supported P.H., and J.A.M.

Acknowledgments: We would like to thank the numerous volunteers and field assistants who spent months measuring vegetation under difficult field conditions. We would like to offer a special thank you for support in the field to Kelley A Crews-Meyer and our field assistant Gully. Finally, we would like to thank editors and the anonymous reviewers for their comments and suggestions.

Conflicts of Interest: The authors declare no conflict of interest.

Appendix A Derived Burned Area Product

The MODIS MCD64A1 (LP DAAC 2010b) burned area product as described by Giglio et al. (2009) was used to determine the fire locations and fire frequency across Botswana from year 2001 to 2011. Fire frequency was calculated on a per pixel basis based on MODIS derived fire events, with frequency ranging from no fire to a maximum of eleven fires recorded for the northern part of the country along the northern border with Namibia. Due to fire frequency having more influence on vegetation than time since last fire, the estimated uncertainty in date of burn was not incorporated in the fire frequency product as this does not impact the output. Figure A1 shows that most areas that burned have a fire frequency between one fire and three fires for the observed time frame. Most of these areas are located in the northern and central parts of the region. Across all field sites, fire frequency ranged from 0–6, with a total of eight sites having been affected at least once during the time period. Return intervals for fire occurrences seem to be higher at the northern site locations (sites 1–6) while all central to southern sites (sites 7–15) with the exception the Bere and Tshane South (site 10 and 12) were unaffected by fire.

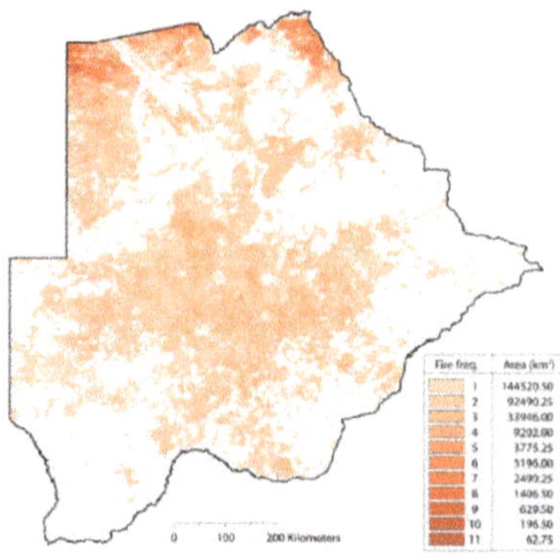

Figure A1. Fire frequency and burned area across Botswana from 2001 to 2011.

References

1. Chapin, S., III; Kofinas, G.P.; Folk, C. *Principles of Ecosystem Stewardship: Resilience-Based Natural Resource Management in a Changing World*; Springer Science and Business: New York, NY, USA, 2009.
2. Scholes, R.; Archer, S. Tree-grass interaction in savannas 1. *Annu. Rev. Ecol. Syst.* **1997**, *28*, 517–544. [CrossRef]
3. Grossman, D.; Gandar, M. Land transformation in South African savanna regions. *S. Afr. Geogr. J.* **1989**, *71*, 38–45. [CrossRef]
4. Acemoglu, D.; Johnson, S.; Robinson, J. An African Success Story: Botswana. CEPR Discussion Paper No 3219. 2002. Available online: https://ssrn.com/abstract=304100 (accessed on 23 July 2013).
5. Robinson, J.A.; Parsons, Q.N. State formation and governance in Botswana. *J. Afr. Econ.* **2006**, *15*, 100–140. [CrossRef]
6. Moleele, N.; Ringrose, S.; Matheson, W.; Vanderpost, C. More woody plants? The status of bush encroachment in Botswana's grazing areas. *J. Environ. Manag.* **2002**, *64*, 3–11. [CrossRef]
7. Ringrose, S.; Matheson, W.; Wolski, P.; Huntsman-Mapila, P. Vegetation cover trends along the Botswana Kalahari transect. *J. Arid Environ.* **2003**, *54*, 297–317. [CrossRef]
8. Archer, S.R. Have southern Texas savannas been converted to woodlands in recent history? *Am. Nat.* **1989**, *134*, 545–561. [CrossRef]
9. Van Auken, O. Shrub invasions in North American semiarid grasslands. *Annu. Rev. Ecol. Syst.* **2000**, *31*, 197–215. [CrossRef]
10. Eldridge, D.J.; Bowker, M.A.; Maestre, F.T.; Roger, E.; Reynolds, J.F.; Whitford, W.G. Impacts of shrub encroachment on ecosystem structure and functioning: Towards a global synthesis. *Ecol. Lett.* **2011**, *14*, 709–722. [CrossRef]
11. Sankaran, M.J.; Ratnam, J.; Hanan, N.P. Woody cover in African savannas: The role of resources, fire, and herbivory. *Glob. Ecol. Biogeogr.* **2008**, *17*, 236–245. [CrossRef]
12. Wigley, B.J.; Bond, W.J.; Hoffman, M. Thicket expansion in a South African savanna under divergent land use: Local vs global drivers? *Glob. Chang. Biol.* **2010**, *16*, 964–976. [CrossRef]
13. O'Connor, T.G.; Puttick, J.R.; Hoffman, M.T. Bush encroachment in southern Africa: Changes and causes. *Afr. J. Range Forage Sci.* **2014**, *31*, 67–88. [CrossRef]
14. Stevens, N.; Lehmann, C.E.R.; Murphy, B.P.; Durigan, G. Savanna woody encroachment is widespread across three continents. *Glob. Chang. Biol.* **2017**, *23*, 235–244. [CrossRef]
15. Archer, S.R.; Andersen, E.M.; Predick, K.I.; Schwinning, S.; Steidl, R.J.; Woods, S.R. Woody Plant Encroachment: Causes and Consequences. In *Rangeland Systems*; Briske, D., Ed.; Springer Series on Environmental Management; Springer: Cham, Switzerland, 2017; pp. 25–84.
16. Sankaran, M.J.; Hanan, N.P.; Scholes, R.J.; Ratnam, J.; Augustine, D.J.; Cade, B.S.; Gignoux, J.; Higgins, S.I.; Le Roux, X.; Ludwig, F.; et al. Determinants of woody cover in African savannas. *Nature* **2005**, *438*, 846. [CrossRef]
17. Ward, D. Do we understand the causes of bush encroachment in Africa savannas? *Afr. J. Range Forage Sci.* **2005**, *22*, 101–105. [CrossRef]
18. Fensham, R.J.; Fairfax, R.J.; Archer, S.R. Rainfall, land use and woody vegetation cover change in semi-arid Australian savanna. *J. Ecol.* **2005**, *93*, 596–606. [CrossRef]
19. Kraaij, T.; Ward, D. Effects of rain, nitrogen, fire and grazing on tree recruitment and early survival in bush-encroached savanna, South Africa. *Plant Ecol.* **2006**, *186*, 235–246. [CrossRef]
20. Yang, X.; Crews, K.A.; Yan, B. Analysis of the pattern of potential woody cover in Texas savanna. *Int. J. Appl. Earth Obs. Geoinform.* **2016**, *52*, 527–531. [CrossRef]
21. Joubert, D.F.; Smit, G.N.; Hoffman, M.T. The influence of rainfall, competition and predation on seed production, germination and establishment of an encroaching Acacia in an arid Namibian savanna. *J. Arid Environ.* **2013**, *91*, 7–13. [CrossRef]
22. Kulmatiski, A.; Beard, K.H. Woody plant encroachment facilitated by increased precipitation intensity. *Nat. Clim. Chang.* **2013**, *3*, 833. [CrossRef]
23. Mistry, J.; Beraldi, A. *World Savannas: Ecology and Human Use*; Routledge: London, UK, 2014.
24. Seymour, C.L. Grass, rainfall and herbivores as determinants of Acacia eriobola (Meyer) recruitment in an African savanna. *Plant Ecol.* **2008**, *197*, 131–138. [CrossRef]

25. Walter, H. Die Verbuschung, eine Ercheinung der subtropischen savannengebiete, und ihre ökologischen ursachen. *Vegetatio* **1954**, *5*, 6–10. [CrossRef]

26. Skarpe, C. Shrub layer dynamics under different herbivore densities in arid savanna, Botswana. *J. Appl. Ecol.* **1990**, *27*, 873–885. [CrossRef]

27. Perkins, J.; Thomas, D. Spreading deserts or spatially confined environmental impacts? Land degradation and cattle ranching in the Kalahari desert of Botswana. *Land Degrad. Dev.* **1993**, *4*, 179–194. [CrossRef]

28. Holdo, R.M. Revisiting the two-layer hypothesis: Coexistence of alternative functional rooting strategies in savannas. *PLoS ONE* **2013**, *8*, e69625. [CrossRef]

29. Brown, J.R.; Archer, S. Shrub invasion of grassland: Recruitment is continuous and not regulated by herbaceous biomass or density. *Ecology* **1999**, *80*, 2385–2396. [CrossRef]

30. Wiegand, K.; Ward, D.; Saltz, D. Multi-scale patterns and bush encroachment in an arid savanna with a shallow soil layer. *J. Veg. Sci.* **2005**, *16*, 311–320. [CrossRef]

31. Angassa, A.; Oba, G. Herder perceptions on impacts of range enclosures, crop farming, fire ban and bush encroachment on the rangelands of Borana, southern Ethiopia. *Hum. Ecol.* **2008**, *36*, 201–215. [CrossRef]

32. Eldridge, D. Trampling of microphytic crusts on calcareous soils, and its impact on erosion under rain-impacted flow. *Catena* **1998**, *33*, 221–239. [CrossRef]

33. Thomas, A.D. Impact of grazing intensity on seasonal variations in soil organic carbon and soil CO_2 efflux in two semiarid grasslands in southern Botswana. *Philos. Trans. R. Soc. B Biol. Sci.* **2012**, *367*, 3076–3086. [CrossRef]

34. Dougill, A.J.; Thomas, A.D. Kalahari sand soils: Spatial heterogeneity, biological soil crusts and land degradation. *Land Degrad. Dev.* **2004**, *15*, 233–242. [CrossRef]

35. Wang, L.; D'Odorico, P.; Manzoni, S.; Porporato, A.; Macko, S. Soil carbon and nitrogen dynamics in southern African savannas: The effect of vegetation-induced patch-scale heterogeneities and large scale rainfall gradients. *Clim. Chang.* **2009**, *94*, 63–74. [CrossRef]

36. O'Halloran, L.R.; Shugart, H.H.; Wang, L.; Caylor, K.K.; Ringrose, S.; Kgope, B. Nutrient limitations on aboveground grass production in four savanna types along the Kalahari Transect. *J. Arid Environ.* **2010**, *74*, 284–290. [CrossRef]

37. Dougill, A.J.; Thomas, D.S.; Heathwaite, A.L. Environmental change in the Kalahari: Integrated land degradation studies for nonequilibrium dryland environments. *Ann. Assoc. Am. Geogr.* **1999**, *89*, 420–442. [CrossRef]

38. Berkeley, A.; Thomas, A.D.; Dougill, A.J. Cyanobacterial soil crusts and woody shrub canopies in Kalahari rangelands. *Afr. J. Ecol.* **2005**, *43*, 137–145. [CrossRef]

39. Stringer, L.; Reed, M. Land degradation assessment in southern Africa: Integrating local and scientific knowledge bases. *Land Degrad. Dev.* **2007**, *18*, 99–116. [CrossRef]

40. Higgins, S.I.; Bond, W.J.; Trollope, W.S. Fire, resprouting and variability: A recipe for grass-tree coexistence in savanna. *J. Ecol.* **2000**, *88*, 213–229. [CrossRef]

41. van Wilgen, B.W.; Trollope, W.S.; Biggs, H.C.; Potgieter, A.; Brockett, B.H. Fire as a driver of ecosystem variability. In *The Kruger Experience: Ecology and Management of Savanna Heterogeneity*; Island Press: Washington, DC, USA, 2003.

42. Sankaran, M.J.; Ratnam, J.; Hanan, N.P. Tree-grass coexistence in savannas revisited—Insights from an examination of assumptions and mechanism invoked in existing models. *Ecol. Lett.* **2004**, *7*, 480–490. [CrossRef]

43. Trollope, W. Controlling bush encroachment with fire in the savanna areas of South Africa. *Proc. Annu. Congr. Grassl. Soc. S. Afr.* **1980**, *15*, 173–177. [CrossRef]

44. Van der Walt, P.T.; Le Riche, E.A.N. The influence of veld fire on an Acacia eriobola community in the Kalahari Gemsbok National Park. *Koedoe* **1984**, *27*, 103–106. [CrossRef]

45. Seymour, C.L.; Huyser, O. Fire and the demography of camelthorn (Acacia eriobola Meyer) in the southern Kalahari—Evidence for a bonfire effect? *Afr. J. Ecol.* **2008**, *46*, 594–601. [CrossRef]

46. Bond, W.; Midgley, G.; Woodward, F. The importance of low atmospheric CO_2 and fire in promoting the spread of grasslands and savannas. *Glob. Chang. Biol.* **2003**, *9*, 973–982. [CrossRef]

47. Mouillot, F.; Field, C.B. Fire history and the global carbon budget: A 1x1 fire history reconstruction for the 20th Century. *Glob. Chang. Biol.* **2005**, *11*, 398–420. [CrossRef]

48. Joubert, D.F.; Smit, G.N.; Hoffman, M.T. The role of fire in preventing transitions from a grass dominated state to a bush thickened state in arid savannas. *J. Arid Environ.* **2012**, *87*, 1–7. [CrossRef]

49. Lomolino, M.V. A call for a new paradigm of island biogeography. *Glob. Ecol. Biogeogr.* **2000**, *9*, 1–6. [CrossRef]

50. Marquet, P.A.; Fernández, M.; Navarrete, S.A.; Valdovinos, C. Diversity emerging: Toward a deconstruction of biodiversity patterns. In *Frontiers of Biogeography: New Directions in the Geography of Nature*; Lomolino, M.V., Heaney, L.R., Eds.; Sinauer Associates Inc Publishers: Sunderland, MA, USA, 2004; pp. 191–209.

51. Joubert, D.F.; Rothauge, A.; Smit, G.N. A conceptual model of vegetation dynamics in the semiarid Highland savanna of Namibia, with particular reference to bush thickening by Acacia mellifera. *J. Arid Environ.* **2008**, *72*, 2201–2210. [CrossRef]

52. Hijmans, R.J.; Cameron, S.E.; Parra, J.L.; Jones, P.G.; Jarvis, A. Very high resolution interpolated climate surfaces for global land areas. *Int. J. Climatol.* **2005**, *25*, 1965–1978. [CrossRef]

53. Krebs, C.J. *Ecological Methodology*; Addison Welsey Educational Publishers: Menlo Park, CA, USA, 1999.

54. Palgrave, K.C. *Trees of Southern Africa*; Struik, C., Ed.; Penguin Random House South Africa: Cape Town, South Africa, 1977.

55. Meyer, T.; D'Odorico, P.; Okin, G.S.; Shugart, H.H.; Caylor, K.K.; O'Donnell, F.C.; Bhattachan, A.; Dintwe, K. An analysis of structure: Biomass structure relationships for characteristic species of the western Kalahari, Botswana. *Afr. J. Ecol.* **2013**, *52*, 20–29. [CrossRef]

56. DSM. Digital Atlas of Botswana. In *DOSA Mapping*; Gabarone, Botswana, 2003.

57. Giglio, L.; Loboda, T.; Roy, D.P.; Quayle, B.; Justice, C.O. An active-fire based burned area mapping algorithm for the MODIS sensor. *Remote Sens. Environ.* **2009**, *113*, 408–420. [CrossRef]

58. Land Processing Distributed Active Archive Center (LP DAAC). MODIS/Terra and Aqua Burned Area Monthly L3 Global 500m SIN Grid V006, NASA EOSDIS Land Processes DAAC, USGS/Earth Resources Observation and Science (EROS) Center, Siox Falls, South Dakota (https://lpdaac.usgs.gov). Available online: https://lpdaac.usgs.gov/datasetdiscovery/modis/modisproductstable/mcd64a1v006 (accessed on 23 January 2018).

59. Craig, G.C. *Department of Wildlife and National Parks. Animal Distribution, Numbers and Trends in the Kalahari Ecosystem 1989–2005*; Department of Wildlife and National Parks: Gaborone, Botswana, 2010.

60. Department of Water Affairs. *Boreholes*; Department of Water Affairs: Maun, Botswana, 2000.

61. R Core Team. *R: A Language and Environment for Statistical Computing*; R Foundation for Statistical Computing: Vienna, Austria, 2013; Available online: https://www.R-project.org/ (accessed on 1 February 2019).

62. Peterson, E. Estimating cover of an invasive grass (Bromus tectorum) using tobit regression and phenology derived from two dates of Landsat ETM+ data. *Int. J. Remote Sens.* **2005**, *26*, 2491–2507. [CrossRef]

63. Osgood, D.W.; Finken, L.L.; McMorris, B.J. Analyzing multiple-item measures of crime and deviance II: Tobit regression analysis of transformed scores. *J. Quant. Criminol.* **2002**, *18*, 319–347. [CrossRef]

64. Laue, J.E.; Arima, E.Y. Spatially explicit models of land abandonment in the Amazon. *J. Land Use Sci.* **2016**, *11*, 48–75. [CrossRef]

65. Tobin, J. Estimation of relationships for limited dependent variables. *Econ. J. Econ. Soc.* **1958**, *26*, 24–36. [CrossRef]

66. Henningsen, A. CensReg: Censored Regressions (Tobit) Models. R Package Version 0.5-26. Available online: https://CRAN.R-project.org/package=censReg (accessed on 2 February 2019).

67. Henningsen, A.; Toomet, O. maxLik; A package for maximum likelihood estimation in R. *Comput. Stat.* **2011**, *26*, 443–458. [CrossRef]

68. White, M.A.; Asner, G.P.; Nemani, R.; Privette, J.L.; Running, S.W. Measuring fractional cover and leaf area index in arid ecosystems: Digital camera, radiation transmittance and laser altimetry methods. *Remote Sens. Environ.* **2000**, *74*, 45–57. [CrossRef]

69. Ward, D.; Esler, K.J. What are the effects of substrate and grass removal on recruitment of Acacia mellifera seedlings in a semi-arid environment. *Plant Ecol.* **2011**, *212*, 245–250. [CrossRef]

70. Le Houérou, H.N.; Corra, M. Some browse plants of/ethiopia. In *Browse in Africa*; Le Houerou, H.N., Ed.; ILCA: Addis Abada, Ethiopia, 1980; pp. 109–114.

71. Moleele, N.; Perkins, J.S. Encroaching woody plant species and boreholes: Is cattle density the main driving factor in the Olifants Drift communal grazing lands, south-eastern Botswana? *J. Arid Environ.* **1998**, *40*, 245–253. [CrossRef]

72. Dublin, H.T.; Sinclair, A.R.E.; McGlade, J. Elephants and fire as causes of multiple stable states in the Serengeti-Mara woodlands. *J. Anim. Ecol.* **1990**, *59*, 1147–1164. [CrossRef]

73. Anderies, J.M.; Janssen, M.A.; Walker, B.H. Grazing, management, resilience and the dynamics of fire-driven rangeland system. *Ecosystems* **2002**, *5*, 23–44. [CrossRef]

74. D'Odorico, P.; Laio, F.; Ridolfi, L. A probabilistic analysis of fire-induced tree-grass coexistence in savannas. *Am. Nat.* **2006**, *167*, E79–E87. [CrossRef]

75. Twomey, M.; Brodte, E.; Ute, J.; Brose, U.; Crowe, T.P.; Emmerson, M.C. Idiosyncratic species effects confound size-based predictions of responses to climate change. *Philos. Trans. R. Soc. B Biol. Sci.* **2012**, *367*, 2971–2978. [CrossRef]

76. Mutuku, P.M.; Kenfack, D. Effect of local topographic heterogeneity on tree species assembly in an Acacia-dominated African savanna. *J. Trop. Ecol.* **2019**, *35*, 46–56. [CrossRef]

77. Joubert, D.F.; Zimmermann, I. The potential impacts of wood harvesting of bush thickening species on biodiversity and ecological processes. *Proc. Natl. For. Res. Work. Wind. Namibia* **2002**, *2*, 67–98.

78. Legendre, P. Spatial Autocorrelation: Trouble or New Paradigm? *Ecology* **1993**, *74*, 1659–1673. [CrossRef]

79. Keitt, T.H.; Bjornstad, O.N.; Dixon, P.M.; Citron-Pousty, S. Accounting for spatial pattern when modeling organism-environmental interactions. *Ecography* **2002**, *25*, 616–625. [CrossRef]

80. Miller, J.A.; Franklin, J.; Aspinall, R. Incorporating spatial dependence in predictive vegetation models. *Ecol. Model.* **2007**, *202*, 225–242. [CrossRef]

81. Holloway, P.; Miller, J.A. Exploring Spatial Scale, Autocorrelation and Nonstationarity of Bird Species Richness Patterns. *Int. J. Geo Inf.* **2015**, *4*, 783–798. [CrossRef]

 *land*

Article

Habitat Climate Change Vulnerability Index Applied to Major Vegetation Types of the Western Interior United States

Patrick J. Comer [1,*], Jon C. Hak [1], Marion S. Reid [1], Stephanie L. Auer [2], Keith A. Schulz [1], Healy H. Hamilton [2], Regan L. Smyth [2] and Matthew M. Kling [3]

[1]   NatureServe, 1680 38th Street, Suite 120, Boulder, CO 80301, USA
[2]   NatureServe, 2511 Richmond Highway, Suite 930, Arlington, VA 22202, USA
[3]   Department of Integrative Biology, University of California, Berkeley, CA 94720, USA
*   Correspondence: pcomer0318@gmail.com; Tel.: +1-703-797-4802

Received: 20 May 2019; Accepted: 3 July 2019; Published: 6 July 2019

Abstract: We applied a framework to assess climate change vulnerability of 52 major vegetation types in the Western United States to provide a spatially explicit input to adaptive management decisions. The framework addressed climate exposure and ecosystem resilience; the latter derived from analyses of ecosystem sensitivity and adaptive capacity. Measures of climate change exposure used observed climate change (1981–2014) and then climate projections for the mid-21st century (2040–2069 RCP 4.5). Measures of resilience included (under ecosystem sensitivity) landscape intactness, invasive species, fire regime alteration, and forest insect and disease risk, and (under adaptive capacity), measures for topo-climate variability, diversity within functional species groups, and vulnerability of any keystone species. Outputs are generated per 100 km² hexagonal area for each type. As of 2014, moderate climate change vulnerability was indicated for >50% of the area of 50 of 52 types. By the mid-21st century, all but 19 types face high or very high vulnerability with >50% of the area scoring in these categories. Measures for resilience explain most components of vulnerability as of 2014, with most targeted vegetation scoring low in adaptive capacity measures and variably for specific sensitivity measures. Elevated climate exposure explains increases in vulnerability between the current and mid-century time periods.

Keywords: adaptive capacity; climate change vulnerability; exposure; resilience; sensitivity; vegetation

---

1. Introduction

Climate change represents a globally pervasive stress on natural ecosystems. Temperature and precipitation regimes drive ecosystem productivity and natural dynamics, such as the rate of plant growth, the frequency of natural wildfire, and seasonal streamflow [1]. Paleoecological research has shown that past episodes of climate change triggered transformation of natural communities at regional and local scales with varying speed and magnitude [2,3]. As the rate of climate change increases, substantial shifts in key ecological processes are likely to cascade through natural communities, resulting in altered productivity, change in species composition, local extinctions, and many instances of ecological degradation or collapse [4].

Conservation practitioners often lack a sufficient understanding of the many linkages between changing climate and key ecological processes. Nor do they fully understand the many interactions of climate-induced stress with other ecological stressors, such as those tied to land use, which may have already reduced the resiliency of many natural communities [5]. However, because of the controlling link between climate and many ecological processes, and the individualistic responses of component species, natural communities could transform in unprecedented ways [6,7]. Therefore, in any given place, a transparent assessment of climate change vulnerability for natural communities is needed to help quantify risk of ecological degradation or collapse.

A repeatable and transparent index of climate change vulnerability designed for natural communities helps to determine those types that, in all or part of their distribution, are most at risk of climate change impacts. It can provide an early warning of elevated risk for associated species, and a baseline for developing scientifically grounded, ecosystem-based strategies for climate change adaptation. This Habitat Climate Change Vulnerability Index (HCCVI) presented in this paper integrates variables from other assessments and results in products that support practical decision making and communication.

Vulnerability Assessment at Different Levels of Ecological Organization

Climate change vulnerability assessments may address different levels of ecological organization, such as species, communities, or landscapes. The species level is the most common focus for vulnerability assessment and consequently has received extensive attention in the literature [8–11]. Based in autecology, trait-based approaches examine projected climate change where the species occurs, aspects of the genetic variation, natural history, physiology, and landscape context to assess sensitivity and adaptive capacity [12].

Assessments of landscapes often produce spatially explicit results for interpretation at regional scales. Evaluation of exposure may result in maps showing where climate stress is indicated to be greatest, whereas examination of the potential climate-change effects on disturbance regimes or invasive species can address aspects of sensitivity [13–15]. Adaptive capacity can be measured through examination of the heterogeneity of topography, moisture gradients, or microclimates under the assumption that more diverse landscapes provide more opportunities for organisms to find climate refugia than homogeneous ones [16].

Assessing the vulnerability of natural community types can provide a useful complement to both landscape and species assessments. Whereas landscape assessments indicate a high potential for climate-change impacts in certain regions, analysis of component communities is based on synecology and aims to more directly measure how climate change will impact species assemblages, ecological processes, structure and function, and is a next logical step to identify practical adaptation strategies where local management of vegetation is a common form of resource management [7].

Few examples exist for assessing vulnerability at the community level of organization. In two examples, they assessed vulnerability of European forests [17], and coastal communities [18], but both took conceptual approaches based on these socio-ecological systems at continental scales. They were concerned mainly about ecosystem outputs of goods and services and aiming to inform societal responses in forestry and fisheries sectors.

In one recent example focusing more squarely at natural community types themselves [19], 31 major vegetation types in California were assessed considering climate projections by the year 2100, sensitivity and adaptive capacity estimates of dominant species for each type, and spatial disruption that might inhibit species movement over time. They used a basin characterization model [20] to express climate change exposure. Working at a global scale, assessment for terrestrial ecosystems [21] emphasized the need to address all three factors of exposure, sensitivity, and adaptive capacity.

Here, we demonstrate a practical framework for climate change vulnerability assessment, focusing on natural community types themselves. This current assessment builds upon prior efforts [22] to address major vegetation types, such as sagebrush shrubland and pinyon-juniper woodlands [23], with an initial focus on types that dominate lands managed by the Bureau of Land Management (BLM) in the conterminous Western US. Several of these types also have extensive distributions in neighboring Mexico or Canada. Below we explain our methods with the outputs of each step illustrated with one common sagebrush shrubland type, and then summarize findings and discuss implications for all 52 types that extend over 3.2 million km^2 of Western North America.

We integrate measures of climate exposure with a series of measures for ecosystem resilience. Our intent is to complete analysis of terrestrial ecosystems or natural community types conceptualized at relatively local scales. Our approach facilitates the accumulation of results for many ecosystem types occurring across countries and continents. Results from this form of ecologically based analysis can then be applied to climate change adaptation in a variety of socio-ecological contexts across the range of distribution of the natural community type.

2. Materials and Methods

2.1. Analytical Framework for Vulnerability Assessment

This index approach to vulnerability assessment aims to organize a series of sub-analyses in a coherent structure that will shed light on distinct components of vulnerability, so that each can be evaluated individually, or in combination. Our approach parallels related indexing of climate change vulnerabilities for species [24]. The components of climate change vulnerability are organized into primary categories of Exposure and Resilience. Resilience is further subdivided into subcategories of Sensitivity and Adaptive Capacity (Figure 1). For the HCCVI, these terms are defined as follows:

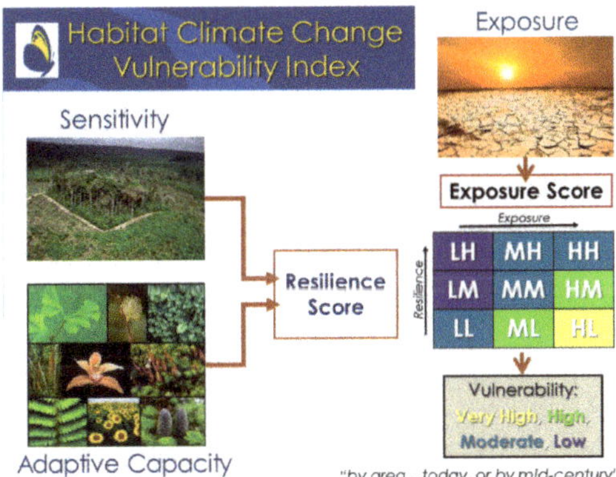

Figure 1. Analytical framework for the habitat climate change vulnerability index

Exposure refers to the rate, magnitude, and nature of climate-induced stress on the community. Exposure encompasses trends in climate, such as changes in temperature and precipitation regimes, and any predicted effects on ecosystem-specific processes. Analyses of exposure consider change in climate variables themselves, or if possible, their resulting effects that cause increasing ecosystem stress, changing dynamic processes such as wildfire or hydrological regimes, or coastal dynamics. This definition of exposure aligns closely with other common applications to climate change vulnerability assessments where changing climate is viewed as an extrinsic driver or pressure on the system of interest [12].

Resilience encompasses intrinsic factors that are likely to affect ecological responses of the natural community to changing climate. These include factors commonly described for ecological resilience, as defined by Holling [25] and Gunderson [26]. Walker et al. [27] defined ecological resilience as "the capacity of a system to absorb disturbance and reorganize while undergoing change and while still retaining essentially the same function, structure, identity, and feedbacks." Under the broader measure of resilience, we include subcategories of Sensitivity and Adaptive Capacity.

- *Sensitivity* focuses on ecosystem stressors that are likely to affect ecological responses of the natural community to climate change. These emphasize human alterations to characteristic patterns and process that can be readily measured across the range of the community type, such as landscape fragmentation, effects of invasive species, or human alterations to other dynamic processes. These alterations are considered independent of climate change, but once identified, are likely to interact with changing climate.

- *Adaptive Capacity* includes natural characteristics that affect the potential for a natural community to cope with climate change. Analyses might consider the natural geophysical variability in climate for the type's distribution. They might also consider functional roles that characteristic species play, such as the relative vulnerabilities to climate change of individual species that provide "keystone" functions, and relative taxonomic diversity within key functional groups (e.g., C3- vs. C4-dominated communities) that characterize the type.

These definitions differ in part from those provided by the IPCC because here we emphasize synecology. Whereas sensitivity in species vulnerability is based more strongly on life history characteristics, natural communities encompass multiple species, each with differing tolerances. Here our sensitivity definition aims to better account for human alterations that affect the community composition and response to changing climate. Our adaptive capacity definition emphasizes natural characteristics of the community type—both biotic (species composition) and abiotic (geophysical), that may (or may not) contribute to resilience in the face of changing climate.

Drawing inspiration from Magness et al. [28] among others in structuring analyses with a logic model, the index scores combine information in two stages, with the first analyses gauging relative climate change resilience by combining scores from intrinsic factors of sensitivity and adaptive capacity. Climate change exposure and resilience are then considered together (combining extrinsic and intrinsic factors) to arrive at an overall gauge of climate change vulnerability (Figure 1).

The HCCVI uses component indicator values to ultimately arrive at a four-level series of scores, i.e., Very High, High, Moderate, and Low vulnerability (Figure 1). These can be derived from relative measures of both Resilience and Exposure. When using quantitative data for measurement, numerical scores are normalized to a 0.0 to 1 scale, with 0.0 indicating ecologically "least favorable" conditions, and 1 indicating "most favorable" conditions. Quartiles of each continuous measure may be used as a starting point to determine the range falling into each of the Very High–Low categories (e.g., ≥0.75 = Low, 0.5–0.75 = Moderate, 0.25–0.50 = High, and ≤0.25 = Very High overall vulnerability). In this application of the framework, all indicators are weighted equally, and we used an arithmetic mean for their combination. Different circumstances and datasets for component measures, or different needs for reporting, may suggest alternative weightings or thresholds to delineate these categories. See Tonmoy et al. [29] for a review of issues associated with indicator treatment in vulnerability assessments. For this framework, we emphasize the need to standardize indicator data in some form to enable combination with other component analysis results.

Very High climate change vulnerability results from combining high exposure with low resilience (i.e., both trending toward "least favorable" scores). These are circumstances where climate change stress and its effects are expected to be most severe, and relative resilience is lowest. Ecosystem transformation is most likely to occur in these types.

High climate change vulnerability results from combining either high or moderate exposure with low or medium resilience. Under either combination, climate change stress is anticipated to have considerable impact.

Moderate climate change vulnerability results from a variety of combinations for exposure and resilience; initially with circumstances where both are scored as moderate. However, this also results where resilience is scored high, if combined with either high or medium exposure. Where both resilience and exposure are low, some degree of climate change vulnerability remains.

Low climate change vulnerability results from combining low exposure with high resilience (i.e., both trending toward "most favorable" scores). These are circumstances where climate change stress and its effects are expected to be least severe or absent, and relative resilience is highest.

2.2. Spatial and Temporal Dimensions for Documenting Vulnerability

Climate change vulnerability assessments need to be placed within explicit spatial bounds. For this effort, we summarized component measurements by 100 km^2 hexagon for the distribution of each vegetation type, and then further summarized results within Level III of the Commission for Economic Cooperation (CEC) ecoregions [30] for each vegetation type (Figure 2). These ecoregions provide an appropriate and consistent spatial structure to systematically document climate change vulnerability at national or regional scales.

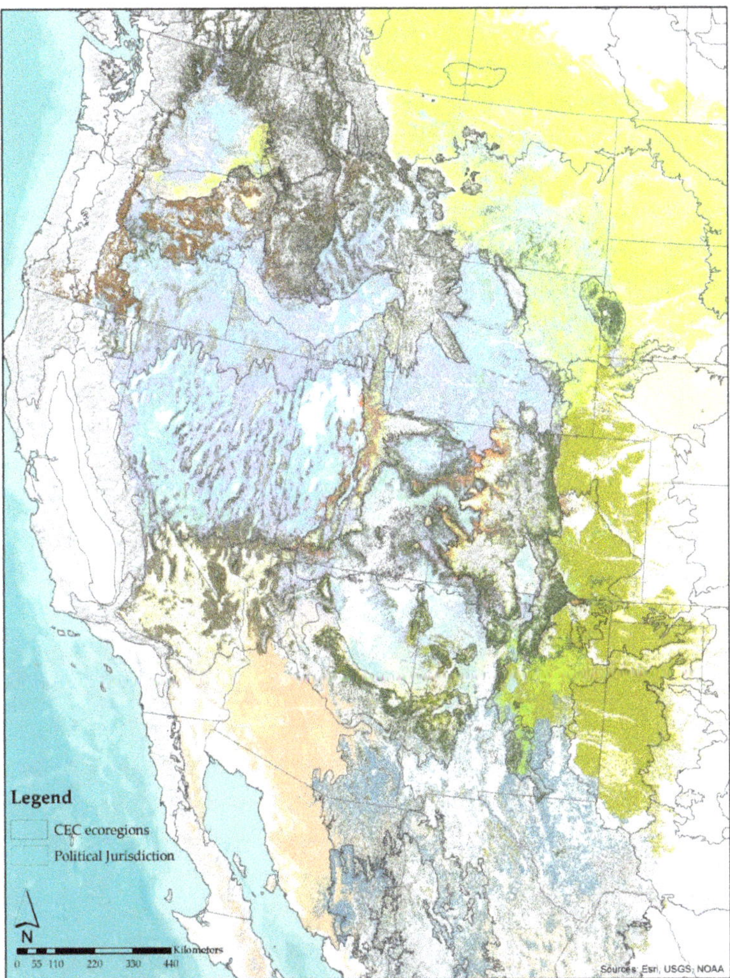

Figure 2. Mapped distributions of 52 ecological system types (too numerous to list here) in Western North America, centered on the USA, with boundaries of the Commission for Economic Cooperation CEC ecoregions [30] used for reporting on vulnerability.

Similarly, climate change vulnerability assessments require a temporal dimension because the magnitude of component measures could vary over time. For this effort, we based initial vulnerability measures on emerging or "current" climate trends using observed climate data (as of 2014). Given both uncertainties associated with climate projections, and the need to apply vulnerability assessments to decisions affecting resources in upcoming decades, climate projections over the upcoming 50-year timeframe (e.g., between 2020 and 2070) provide subsequent realistic timeframes where climate trends can be estimated within acceptable bounds of uncertainty. Therefore, we also assess vulnerability with the mid-21st century timeframe (2040–2069).

2.3. Ecological Classification and Distribution

For this project, we used NatureServe's terrestrial ecological systems classification to define types [31]. The advantage of using this classification system is that it represents an established classification of several hundred upland and wetland types that have been extensively described and mapped by US federal and state resource managers [32,33] and extended into adjacent Canada, Mexico, and across Latin America and the Caribbean [34]. The expected pre-Columbian, or "potential", distribution of each type, mapped at 90 m pixel resolution, was used as the base distribution for assessment (Figure 2). Descriptions of each type can be found at http://explorer.natureserve.org/. We completed additional literature review to further document each type with current knowledge of key ecological processes and stressors, and potential measures for assessing climate change resilience. We completed literature searches using individual communities/systems, common stressors, and functional species groups (e.g., "cool season") as key words using Google Scholar and Colorado State University Library Prospector.

Below, we discuss measures for climate change exposure and resilience applied to each type for this project. Appendix A includes a more detailed explanation of component measures used for ecosystem resilience. For purposes of illustrating our methodology, we will use one example—Intermountain Basins Big Sagebrush Shrubland to depict component steps of the HCCVI framework (Figures 3–7).

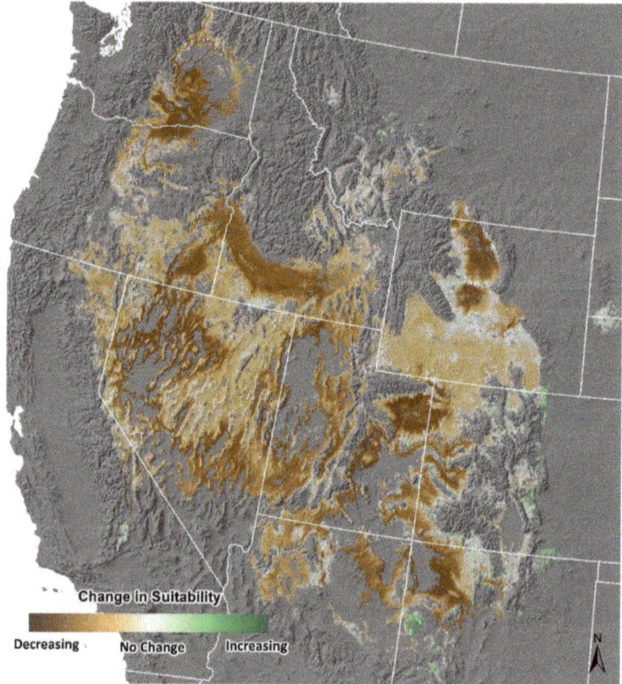

Figure 3. Change in Climate Suitability estimate for 2040–2070 timeframe for Intermountain Basins Big Sagebrush Shrubland.

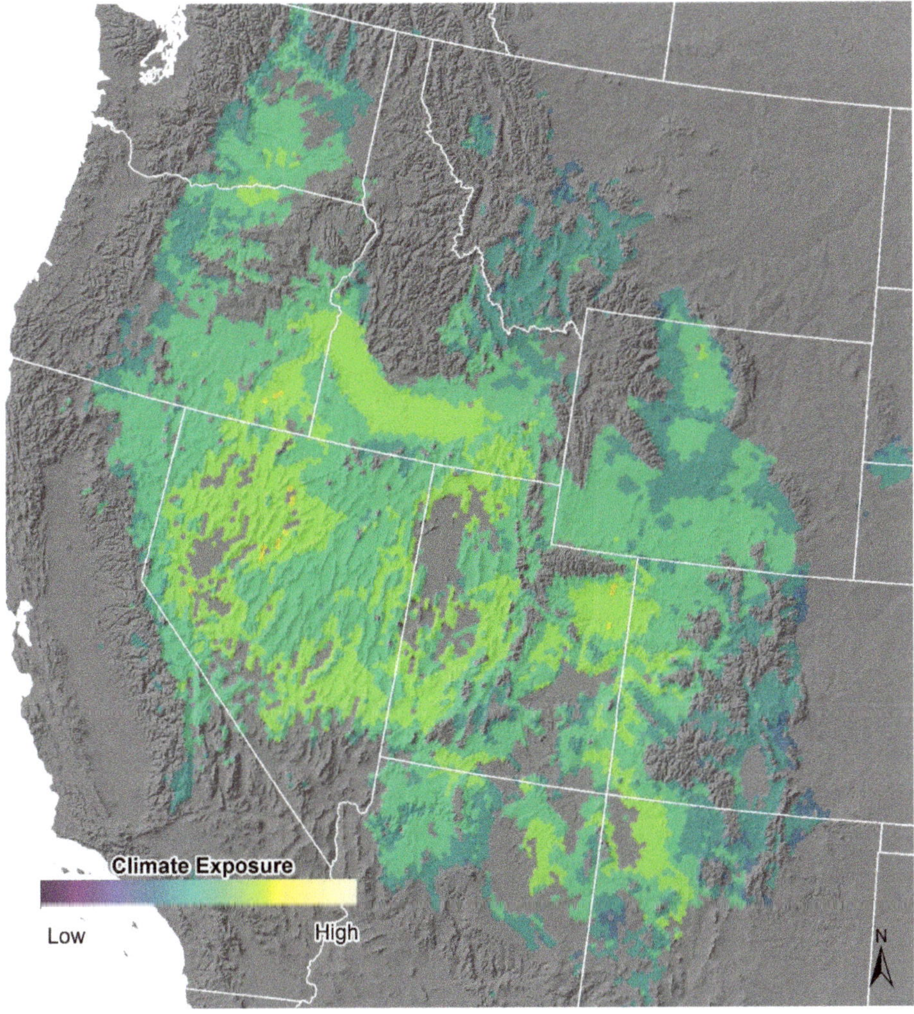

Figure 4. Climate Exposure estimate for 2040–2070 timeframe as index scores (i.e., low–high climate stress) summarized by 100 km² hexagon for Intermountain Basins Big Sagebrush Shrubland.

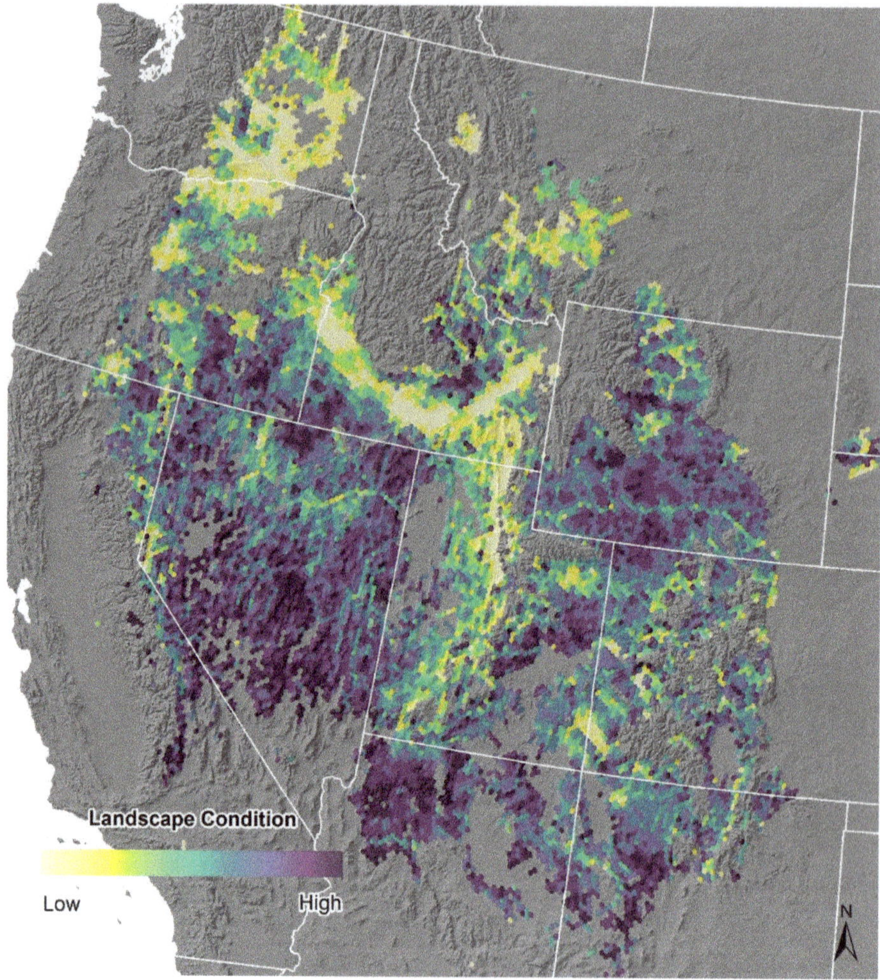

Figure 5. Sensitivity measure of landscape condition and index scores summarized by 100 km^2 hexagon for Intermountain Basins Big Sagebrush Shrubland.

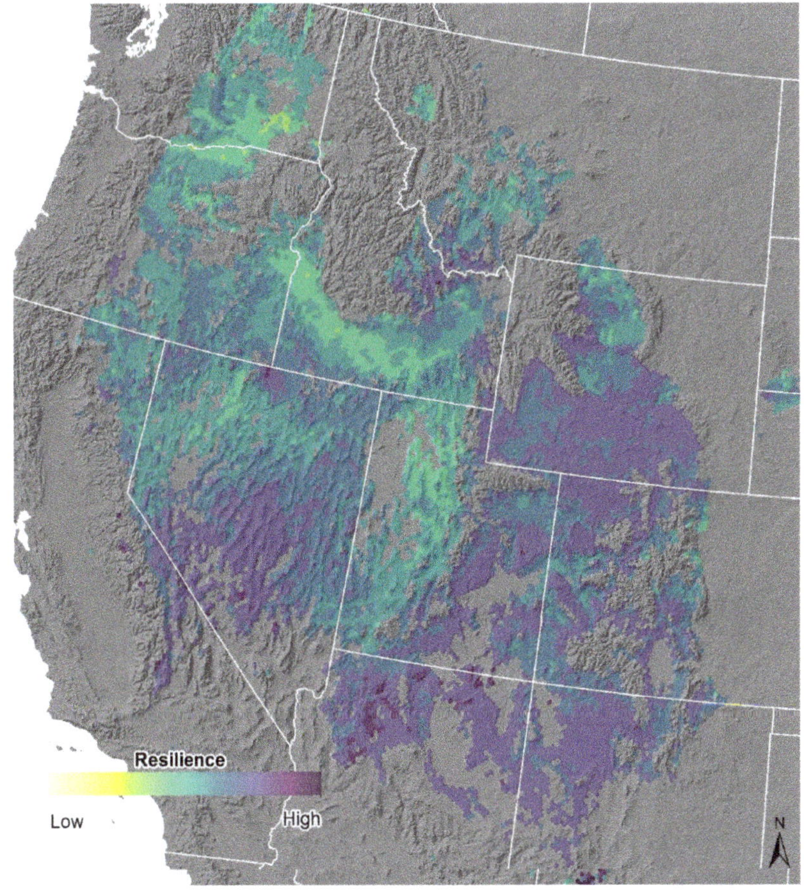

Figure 6. Overall Resilience measure summarized by 100 km^2 hexagon for Intermountain Basins Big Sagebrush Shrubland.

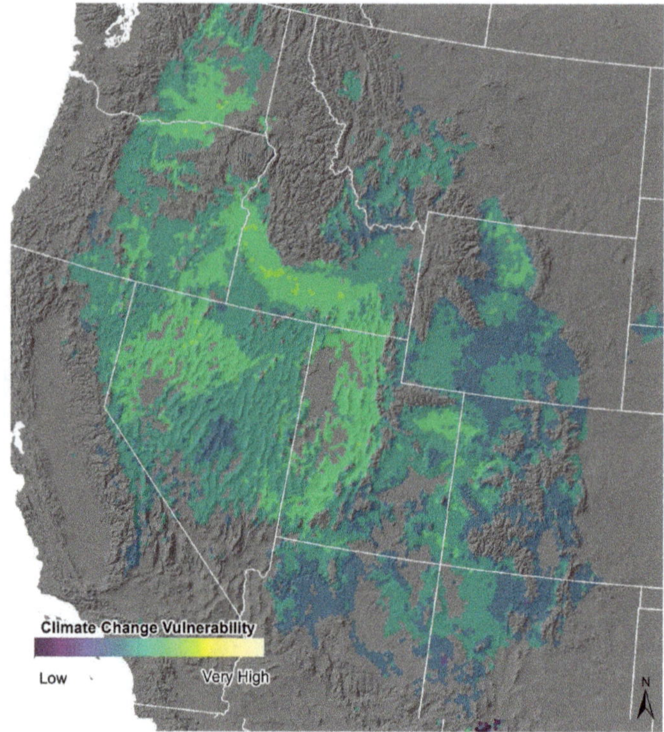

Figure 7. Overall Climate Change Vulnerability estimate for 2040–2070 summarized by 100 km^2 hexagon for Intermountain Basins Big Sagebrush Shrubland.

2.4. Climate Change Exposure

We characterized the baseline climate niche for each vegetation type using historical climate data for the mid-20th century and the potential historical distribution of the type (detail below). This provides a baseline of suitable conditions from which to compare climate trends from subsequent time periods to clarify the significance of measurable change. The entire potential distribution for each community type was used to ensure that the entire range of possible suitable climate conditions was represented for each type. Using distributions that have been affected by land conversion could skew how we define climate suitability for a vegetation type and hence, how we measure climate exposure. Kling et al. [35] provide additional detail on component model design and performance.

For every grid cell of each vegetation type we calculated a composite index of climate change exposure as the sum of two distinct exposure measures: *suitability change*, which quantifies departure from the historical range of spatial climate variability across the geographic range of that vegetation type, and *typicality*, which quantifies departure from the historical range of year-to-year climate variability at a given pixel location. Each component index ranges from 0.0 to 0.5; these are summed to derive the final combined exposure index ranging from 0 (low climate exposure) to 1 (large climate exposure).

Exposure measures were calculated based on changes in 19 bioclimatic variables derived from monthly temperature and precipitation variables [36]. Using a historical baseline period representing typical 20th century climate (1948–1980), we estimated exposure for both recent observed climate change (1981–2014) and projected future change (2040–2069, RCP 4.5). While related regional analyses investigated differences between RCP 4.5 and 8.5 [37], and our methodology could accommodate either here we utilized just RCP 4.5. These calculations were initiated in 2015, and so at that time 2014 was the most recent climate data available. To maximize data quality across the full spatiotemporal extent of our analysis, we combined gridded climate datasets from several sources. Within the contiguous US, for the baseline and recent periods, temperature data were sourced from TopoWx [38] whereas precipitation data were sourced from PRISM version LT71 [39], both at 810 m resolution. Outside the US, and for the future within the US, all variables were based on 1 km^2 data from ClimateNA [40].

The *suitability change* metric uses niche modeling to estimate changes in a site's climatic suitability for a given vegetation type for a given timeframe. For each vegetation type we fit a random forest [41] model using baseline means of the six most important (of 19) bioclimatic variables and 1,000 presence and absence locations sampled from within the rectangular bounding box of the type's geographic extent. The top six differ among types, but six variables consistently explained a high proportion of model variability. Our choice of RF over alternative Generalized Linear Models (GLM), Generalized Additive Models (GAM), and MaxEnt algorithms, as well as our selection of variables for each type, was based on extensive model performance testing using spatial block cross-validation [42] to control the overfitting that comes from spatial non-independence. Variables were selected using recursive feature elimination [43], with six judged as the best empirical tradeoff between performance and parsimony. The fitted models were used to predict suitability for baseline, current, and future periods. Suitability change was then calculated by subtracting suitability measures across these time periods, and then rescaling these differences into a 0–0.5 index with 0 representing increasing or unchanged suitability, and 0.5 representing a large decrease in suitability.

The *typicality* metric compares a site's current or future mean climate to the mean and yearly variation during a baseline period. For a given climate variable, typicality is calculated as the proportion of baseline years whose absolute deviation from the baseline mean (e.g., in degrees C) is larger than the absolute deviation of the recent mean from the baseline mean. A value of 0 indicates no change, while a value of 1 indicates that the recent or future mean climate is more extreme, in either direction, than any individual year in the baseline period. For each vegetation type we calculated typicality for each of the same six climate variables used in the RF models, and then averaged them to derive a final typicality index. Final typicality values were divided by two to rescale them to the 0.0 to 0.5 range.

Typicality (0.0 to 0.5) and Change in Climate Suitability (0.0 to 0.5) measures were then added together to produce an overall exposure measure, resulting in per pixel values within a 0.0–1.0 range, with 1.0 indicating high climate exposure and 0.0 indicating no measurable climate exposure. These per-pixel values were subsequently inverted and averaged for one measure per 100 km^2 hexagon. Figure 3 illustrates one example of mid-21st century estimates of increasing or decreasing climate suitability for one major big sagebrush vegetation type. One can see where most substantial decreases in climate suitability for this type are concentrated in such places as Southeastern Washington, the Snake River Plain of Southern Idaho, Northwestern Nevada, the Big Horn Basin of Wyoming, and the Uinta Basin of Utah.

Figure 4 depicts, for the same type, the result of combining typicality and change in suitability estimates, and then summarizing those per-pixel values to 100 km^2 hexagons. Darker blue areas are forecasted to be least stressed (closer to 1) and yellow areas most stressed (closer to 0.0). This image indicates where a substantial proportion of the range-wide extent scores into the high–very high range of climate exposure.

2.5. Resilience—Ecosystem Sensitivity and Adaptive Capacity

Our measures of Resilience address predisposing conditions—such as extant ecosystem stressors, or natural abiotic or biotic characteristics of the type—that are likely to affect ecological responses of the natural community to changing climate. For example, if exposure measures indicate the need for component species to migrate toward other elevations or latitudes, but the natural landscape is fragmented by intensive land uses, the relative vulnerability of community types in that fragmented landscape could increase [44]. Similarly, the introduction of non-native species may displace native species and/or alter key dynamic processes such as wildfire regimes [45], and both could be exacerbated by climate change. These factors would describe relative climate change sensitivity for a natural community type. This differs in approach from species vulnerability assessments, in that they tend to focus on life-history traits of individual taxa.

Inherent adaptive capacity of natural communities could consider the natural geophysical variability in climate for the type's distribution or the functional roles of species in the community type. Again, climate exposure might indicate a high level of climate stress, natural communities occurring within topographically flat landscapes with potential to retain little variability in microclimates could lack any built-in buffer effects for component species. Likewise, if certain ecosystem functions, such as nitrogen fixation, are limited to one or a few species that characterize the natural community, this could bring additional vulnerability due to the potential for their extirpation over time.

Upon completion of a literature review of each type, and evaluation of available spatial data, we selected four primary indicators suitable for measuring relative sensitivity. To address effects of landscape fragmentation, we used a spatial model for landscape intactness or condition. Since many assessed types are known to be affected by invasive annual grass invasion, a model aiming to measure relative invasion severity was selected. Similarly, since most assessed types have a characteristic natural wildfire regime, a spatial model estimating fire regime departure was used. For forest types, measures of elevated risk from insects or diseases were identified. Three measures of adaptive capacity include both biotic and abiotic factors. Biotics factors included scoring of each type for diversity within identified key functional species groups and relative vulnerability of any identified "keystone" species. One measure of topo-climatic variability was applied for the distribution of each type.

In each of these cases involving spatial models, the model was overlain with the distribution of each vegetation type and scores were transformed to indicate a relative degree of sensitivity or adaptive capacity within a 0.0–1 range, again with 0.0 indicating most severely impacted, or least favorable, conditions while 1 indicating highest integrity, or apparently unaltered, conditions. These scores were each summarized to average values per 100 km^2 hexagon. Figure 5 depicts results from one sensitivity measure of landscape condition for Intermountain Big Sagebrush Shrubland. Darker blue areas indicate apparently least fragmented areas and yellow areas most fragmented. One can see where in substantial portions of this distribution, such as in Eastern Washington, across the Snake River Plain, and south along the Wasatch Front of Utah, this sagebrush type occurs in landscape highly fragmented by intensive land uses.

Again, Appendix A includes a detailed explanation of component measures used for ecosystem resilience.

2.6. Overall Resilience Scores

Overall resilience scores were derived by averaging results for each measure of Sensitivity and for Adaptive Capacity (Figure 1). This combination of up to four Sensitivity measures and up to three Adaptive Capacity measures were averaged together per 100 km^2 hexagon to establish an overall score for resilience. As an example, Figure 6 depicts this overall measure for Intermountain Basins Big Sagebrush Shrubland. One can again see the general pattern of moderate to low resilience in the regional landscapes—such as the Snake River Plain of Southern Idaho—where most intensive land use and effects of invasive plant species are concentrated.

2.7. Overall Climate Change Vulnerability

As noted in Figure 1, the combination of climate change exposure with resilience scores results in the relative vulnerability estimate for a given timeframe. Figure 7 depicts this result as projected for the mid-21st century for Intermountain Basins Big Sagebrush Shrubland. Patterns of vulnerability in this type vary across its distribution, as depicted with 100 km^2 hexagons, and are driven strongly by climate change exposure measures, but are also substantially influenced by component measures of resilience (e.g., Figure 4). While per hexagon outputs are summarized along the 0.0–1 continuum, summary statistics for climate change vulnerability may be desirable with categories expressed as "Very High" "High" Moderate" or "Low" (Figure 1). Here, we used default break-points with quartiles of each continuous measure to determine the range falling into each of the Very High-Low categories (≥0.75 = Low, 0.5–0.75 = Moderate, 0.25–0.50 = High, and ≤0.25 = Very High overall vulnerability). In the case of the Intermountain Basins Big Sagebrush Shrubland, nearly all its distribution is forecasted to fall within the "Moderate" vulnerability as of 2014 and in the "High" range of vulnerability by the mid-21st century.

3. Results

Results for the 52 vegetation types were arranged into 10 categories that reflect major ecological gradients of the region, from high-elevation "Cool Temperate Subalpine Woodlands" down to "Warm Desert Shrublands" (Table 1). This is a high-level summary of analysis scores and overall results for each vegetation type, with proportions of their respective distributions falling in each category (Low–Very High) of vulnerability. On the left are results pertaining to the current timeframe, using climate exposure measure from observed climate trends for the 1981–2014 timeframe. On the right are results using climate exposure measures for the 2040–2069 timeframe.

As of 2014, 50 of 52 types scored within the Moderate vulnerability score for >50% of the rangewide extent. Seven types scored as High vulnerability with >10% of the range-wide extent in this category. No type had >1% of their range scoring in extreme of Very High vulnerability. Of those scoring currently as High Vulnerability, Sonora-Mojave Creosotebush Desert Scrub scored the highest, with 74% of its area scoring as highly vulnerable. Other types currently scoring High are found in warm deserts (Apacherian–Chihuahuan Semi-Desert Grassland and Steppe, Sonoran Paloverde-Mixed Cacti Desert Scrub, Chihuahuan Mixed Desert and Thornscrub, and Western Great Plains temperate shrubland and steppe landscapes (Northwestern Great Plains Mixedgrass Prairie, Rocky Mountain Foothill Limber Pine–Juniper Woodland, Western Great Plains Sand Prairie, Northwestern Great Plains Shrubland) (Table 1).

The primary result of considering projected climate exposure for the mid-21st century was an apparent overall shift in vulnerability scores from Low–Moderate to High ranges (Table 1) for nearly all 52 vegetation types. By the mid-21st century, all but 19 types face high climate change vulnerability with >50% of the area scoring in these categories. This change in overall vulnerability is of course driven by more severe climate exposure measures (into High and Very High categories). Fourteen types scored with over 90% of their distribution in the high vulnerability category (Table 1).

Appendix B provides a more detailed breakdown of the results, summarizing proportional range-wide area by type scoring for climate exposure (both current and mid-21st century timeframes) and for overall resilience and its component scores for sensitivity and adaptive capacity; each with proportions of range-wide extent falling into each of the four binned categories for relative vulnerability. Again, one can see the relative contribution of elevated climate change exposure in the mid-21st century timeframe, with 17 types having over 90% of their range within either the high or very high vulnerability ranges for mid-21st century exposure. These extremes are concentrated in subalpine and montane forests and woodlands, extending down to elevations supporting Ponderosa Pine Woodlands and Savannas. A second concentration of types is in the cool semi-desert shrublands such as xeric sagebrush and Mojave mid-elevation shrubland. Here we also see additional explanation of overall vulnerability scores coming from component measures for resilience. Overall resilience scores were proportionally often in moderate vulnerability, but with 10 types having >50% extent in high vulnerability from their resilience scores. These types are concentrated at lower elevation woodland, shrubland, and grassland types. Overall resilience scores appear to be most strongly driven by specific measures under both sensitivity and adaptive capacity. While most proportional area of each type scored in low to moderate ranges for the combined sensitivity measures, most scored in the high range for adaptive capacity.

Table 1. Terrestrial ecological system types assessed for climate change vulnerability, with *percentage of mapped area* that was scored from Low to Very High Vulnerability using either current climate exposure or those from climate projections to the mid-21st century. Types are organized by elevation-based category and subsequently sorted based on approximate historical extent.

Climate Change Vulnerability	Potential or Historic Extent (Km²)	Current Vulnerability				Mid-21st Century Vulnerability			
		Low	Mod	High	Very High	Low	Mod	High	Very High
		≥0.75	≥0.5 and <0.75	≥0.25 and <0.5	<0.25	≥0.75	≥0.5 and <0.75	≥0.25 and <0.5	<0.25
Terrestrial Ecological System Types		(% Area)	(% Area)	(% Area)	(% Area)	(% Area)	(% Area)	(% Area)	(% Area)
Cool Temperate Subalpine Woodlands									
Northern Rocky Mountain Subalpine Woodland and Parkland	64,645	5%	94%				59%	41%	
Rocky Mountain Lodgepole Pine Forest	16,068	17%	83%	0.05%			21%	79%	0.2%
Aspen & Mountain Mahogany Forests and Woodlands									
Inter-Mountain Basins Aspen-Mixed Conifer Forest and Woodland	27,929	23%	77%				24%	76%	
Rocky Mountain Aspen Forest and Woodland	24,103	20%	80%	0.02%		0.01%	81%	19%	
Inter-Mountain Basins Curl-Leaf Mountain-Mahogany Woodland and Shrubland	7241	53%	47%				32%	68%	
Montane Conifer Forests and Woodlands									
Northern Rocky Mountain Dry-Mesic Montane Mixed Conifer Forest	129,039	5%	95%	0.04%			94%	6%	
Middle Rocky Mountain Montane Douglas-Fir Forest and Woodland	25,585	14%	86%				37%	63%	
Southern Rocky Mountain Dry-Mesic Montane Mixed Conifer Forest and Woodland	15,430	1%	99%				3%	97%	
Southern Rocky Mountain Mesic Montane Mixed Conifer Forest and Woodland	8961	3%	97%				5%	95%	
Ponderosa Pine Woodlands and Savannas									
Northern Rocky Mountain Ponderosa Pine Woodland and Savanna	49,197	11%	79%			0.002%	63%	22%	0.1%
Southern Rocky Mountain Ponderosa Pine Woodland	38,713		98%	2%			62%	38%	
Northwestern Great Plains-Black Hills Ponderosa Pine Woodland and Savanna	18,193		93%	7%		0.3%	100%	0%	
California Montane Jeffrey Pine-(Ponderosa Pine) Woodland	13,264	2%	98%	0.01%			1%	99%	0.3%
Southern Rocky Mountain Ponderosa Pine Savanna	12,892	1%	99%				10%	90%	
East Cascades Oak-Ponderosa Pine Forest and Woodland	470		100%				3%	97%	

Table 1. *Cont.*

Climate Change Vulnerability	Potential or Historic Extent (Km²)	Current Vulnerability				Mid-21st Century Vulnerability			
		Low	Mod	High	Very High	Low	Mod	High	Very High
		≥0.75	≥0.5 and <0.75	≥0.25 and <0.5	<0.25	≥0.75	≥0.5 and <0.75	≥0.25 and <0.5	<0.25
Terrestrial Ecological System Types		(% Area)	(% Area)	(% Area)	(% Area)	(% Area)	(% Area)	(% Area)	(% Area)
Pinyon and Juniper Woodlands									
Madrean Pinyon-Juniper Woodland	44,866	0.04%	97%	3%		0.03%	85%	7%	
Colorado Plateau Pinyon-Juniper Woodland	39,898	0.1%	100%				58%	42%	
Great Basin Pinyon-Juniper Woodland	22,304	0.1%	100%	0.2%			70%	30%	
Southern Rocky Mountain Pinyon–Juniper Woodland	10,466	0.2%	100%				45%	55%	
Southern Rocky Mountain Juniper Woodland and Savanna	10,382		100%				38%	62%	
Columbia Plateau Western Juniper Woodland and Savanna	7784	3%	95%	3%			28%	72%	
Inter-Mountain Basins Juniper Savanna	1531		99%	1%			1%	99%	
Rocky Mountain Foothill Limber Pine-Juniper Woodland	1330	5%	85%	10%			16%	84%	
Colorado Plateau Pinyon-Juniper Shrubland	206		95%	5%			38%	62%	
Cool Temperate Grasslands									
Northwestern Great Plains Mixed Grass Prairie	620,860		68%	32%		3%	74%	24%	
Western Great Plains Sand Prairie	107,307	0.003%	90%	10%			2%	98%	
Columbia Plateau Steppe and Grassland	23,913	0.09%	100%	0.04%			48%	52%	
Inter-Mountain Basins Semi-Desert Grassland	22,532	0.03%	100%	0.07%			11%	89%	
Northern Rocky Mountain Lower Montane-Foothill-Valley Grassland	19,327	1%	97%	1%			67%	33%	
Southern Rocky Mountain Montane-Subalpine Grassland	3054	0.4%	99%	0.2%			3%	97%	

Table 1. *Cont.*

Climate Change Vulnerability	Potential or Historic Extent (Km²)	Current Vulnerability				Mid-21st Century Vulnerability			
		Low	Mod	High	Very High	Low	Mod	High	Very High
		≥0.75	≥0.5 and <0.75	≥0.25 and <0.5	<0.25	≥0.75	≥0.5 and <0.75	≥0.25 and <0.5	<0.25
Terrestrial Ecological System Types		(% Area)	(% Area)	(% Area)	(% Area)	(% Area)	(% Area)	(% Area)	(% Area)
Cool Semi-desert & Temperate Shrubland and Steppe									
Inter-Mountain Basins Big Sagebrush Shrubland	282,439	2%	97%	0.3%		0.01%	73%	27%	
Inter-Mountain Basins Big Sagebrush Steppe	182,114	2%	97%	0.1%			35%	65%	
Inter-Mountain Basins Montane Sagebrush Steppe	83,707	8%	92%	0.010%			70%	30%	
Great Basin Xeric Mixed Sagebrush Shrubland	62,126	7%	93%	0.003%			8%	92%	
Mojave Mid-Elevation Mixed Desert Scrub	54,768	1%	99%	0.3%			19%	81%	
Columbia Plateau Low Sagebrush Steppe	21,361	2%	97%	0.1%			48%	52%	
Rocky Mountain Gambel Oak-Mixed Montane Shrubland	19,637	0.0%	100%	0.01%			3%	96%	0.47%
Northwestern Great Plains Shrubland	9786		70%	30%			36%	63%	0.48%
Colorado Plateau Blackbrush-Mormon-Tea Shrubland	8366		100%				13%	87%	
Rocky Mountain Lower Montane-Foothill Shrubland	8020	0.1%	100%	0.4%			4%	96%	0.002%
Colorado Plateau Mixed Low Sagebrush Shrubland	4250	0.01%	100%	0.4%			6%	94%	
Columbia Plateau Scabland Shrubland	3775		85%	15%			5%	95%	
Wyoming Basins Dwarf Sagebrush Shrubland and Steppe	1107		96%	4%			5%	95%	
Mixed Salt Desert Scrub and Greasewood									
Inter-Mountain Basins Mixed Salt Desert Scrub	95,681	2%	96%	2%			57%	43%	
Inter-Mountain Basins Greasewood Flat	58,269	0.005%	98%	2%			86%	14%	
Inter-Mountain Basins Mat Saltbush Shrubland	10,677		99%	1%			50%	50%	
Warm Temperate Grasslands and Semi-Desert Grasslands									
Western Great Plains Shortgrass Prairie	258,868	13%	87%	0.05%			84%	16%	
Apacherian–Chihuahuan Semi-Desert Grassland and Steppe	249,336		84%	16%		0.001%	59%	41%	
Warm Desert Shrublands									
Sonoran Paloverde-Mixed Cacti Desert Scrub	131,505		63%	37%	0.02%		58%	42%	
Chihuahuan Creosotebush Desert Scrub	92,472		86%	14%			72%	28%	0.05%
Sonora-Mojave Creosotebush-White Bursage Desert Scrub	91,703		26%	74%	0.001%		14%	84%	2%
Chihuahuan Mixed Desert and Thornscrub	21,071		89%	11%			42%	57%	

Components of sensitivity reflect common circumstances for these types found throughout some of the more remote and undeveloped landscape in the Western United States; but that still include substantial areas impacted by past and current ecological stressors. Relative vulnerability contributed by degraded landscape condition varies considerably, with most area by type falling in the low to moderate ranges. For range-wide summary statistics, seven of these vegetation types include >50% of their area in the high to very high vulnerability ranges. Cool temperate grassland types, from the Columbia Plateau east to the Western Great Plains are represented here. These types are found in some of the most intensively cultivated regional landscapes of the types include in this study. Fire regime departure is a substantial contributor to vulnerability, and 16 types include >50% of their area in the high to very high vulnerability ranges. These types tend to be concentrated in lower montane forests and woodlands, cool temperate grasslands, and cool semi-desert shrublands where wildfire suppression policies have resulted in altered successional pathways. Invasive plant models for this analysis primarily pertained to cool desert shrublands and cool temperate grassland types. Several sagebrush and related vegetation types include substantial areas that are indicated as being in the high or very high vulnerability ranges (e.g., Columbia Plateau Western Juniper Woodland and Savanna—54%; Mojave Mid-Elevation Mixed Desert Scrub—33%; Inter-Mountain Basins Big Sagebrush Steppe—30%; Inter-Mountain Basins Big Sagebrush Shrubland—27%), and the effects of invasive plants interact with alterations to wildfire regimes in most of these vegetation types. Vulnerability stemming from forest insect and disease risk applied only to forest and woodland types, and from a range-wide perspective, appear to have contributed less than other factors, but for some types, rather substantial area was measured within the moderate vulnerability range (e.g., Southern Rocky Mountain Mesic Montane Mixed Conifer Forest and Woodland—32%; California Montane Jeffrey Pine-(Ponderosa Pine) Woodland—28%).

Most notably among adaptive capacity measures, the topo-climatic variability is naturally quite low for all but some of the montane forest and woodland types in this region, as most desert shrublands and temperate grasslands occupy vast landscapes of relatively flat to gently rolling topography. Therefore, for most types, most of their area scored within the high to very high vulnerability ranges. Also, for many of these types that dominate the arid interior of Western North America, inherent species diversity is low and recovery from surface disturbance is relatively slow when compared with montane forests and shrublands. Functional species groups identified often centered on nitrogen fixation, soil stability, and other common characteristics of vegetation in semi-arid regions. Relatively low within-type diversity in one or more functional species group led to low scores for this component of climate change vulnerability. As a result, 14 types scored within the very high vulnerability range, while most others scored in the moderate range. Interestingly, only one "keystone" species was identified associated with any of these 52 assessed types. The black-tailed prairie dog (*Cynomys ludovicianus*) was associated with mixed-grass and shortgrass prairie types, but its individual climate change vulnerability was considered to be relatively low.

While the outputs of the HCCVI are most relevant and applicable to analysis of individual ecosystem types, and factors contributing to their relative vulnerability are differentially expressed across their type distribution, one can also combine mapped results for major types to detect patterns of relative vulnerability. Figure 8 depicts the combined results for all 52 types assessed here, by their 100 km^2 hexagon summary unit. Since this set of types represent the predominant vegetation at lower and middle elevations across the region, they can summarize overall patterns of ecosystem type vulnerability for the region. With an extreme of 24 types occurring within a given hexagon, the figure depicts three ways of summarizing vulnerability patterns, using (*left*) the lowest (least vulnerable) scoring type per hexagon, (*right*) the highest (most vulnerable) scoring type, and (*bottom*) the average score of all types present. These views highlight a range of moderate to very high relative vulnerability across key regional landscapes, such as in the Mojave Desert, Columbia Basin, the Great Salt Lake Basin, the Colorado Front Range, the Nebraska Sandhills, and West Texas.

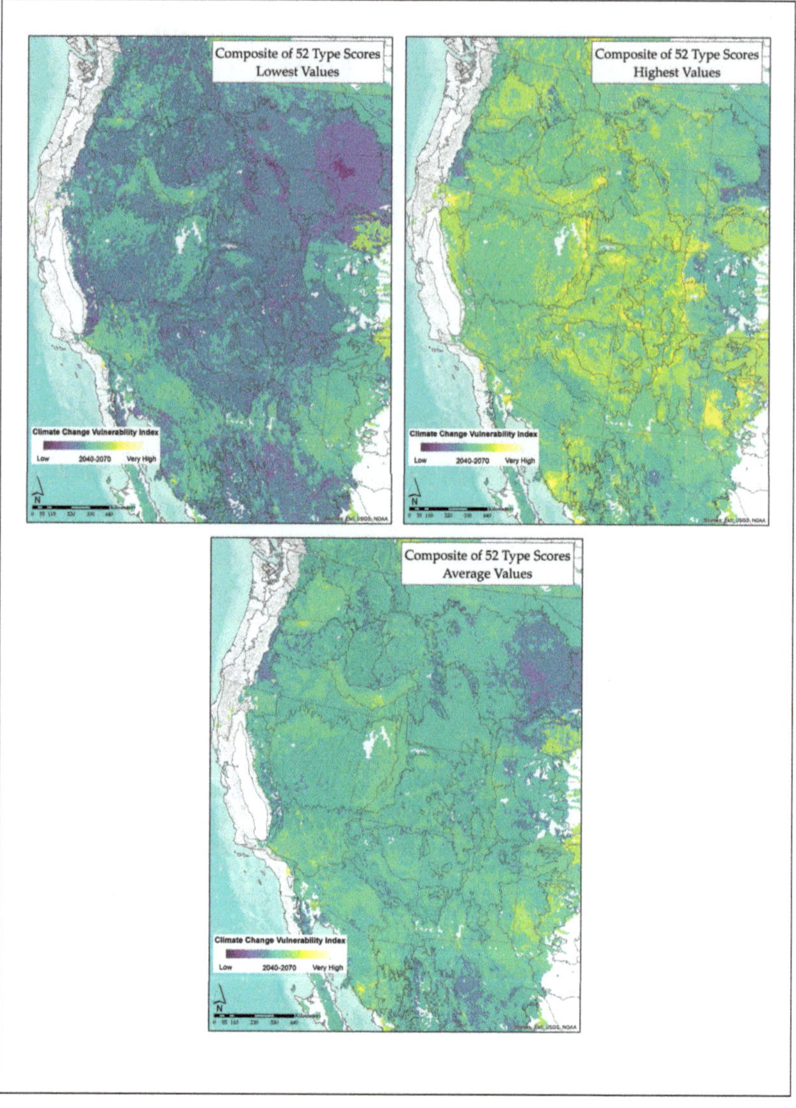

Figure 8. Overall Climate Change Vulnerability estimate for 2040–2070 summarized by 100 km^2 hexagon for all 52 assessed types, with a minimum of one type and a maximum of 24 types occurring within each hexagon, displayed scores include the least, most, and average vulnerability scores (*left to right to bottom*).

4. Discussion

4.1. Vulnerability and Adaptative Vegetation Management

Here we have demonstrated an analytical framework to document relative climate-change vulnerabilities among major upland vegetation types that dominate desert and montane forests of the interior Western USA. While many of these types are well represented in protected and managed public lands [46] that does not shield them from effects of climate change. By integrating available information, we identified types and places where signals of climate change stress are emerging, and where they can be foreseen over upcoming decades. This effort drew inspiration from many similar efforts on the overall structure, measurements used, and available data, but here our analysis is based on synecology and facilitates practical application of results to vegetation at scales commonly addressed in biodiversity conservation and natural resource management. By applying a systematic framework to climate change vulnerability assessment, it generates actionable information targeted to both policy-makers and land managers in support of natural resource conservation decisions.

While traditional natural resource management has tended to be 'retrospective'—utilizing knowledge of past and current conditions to inform today's management actions—conservation professionals are increasingly required to more rigorously forecast future conditions. This forecasting strives to determine the nature and magnitude of change likely to occur, and then translate that knowledge to current decision-making timeframes. It is no longer sufficient to assess "how are we doing?" and then decide what actions should be prioritized for the upcoming 15-year management plan. One must now ask "how is it changing, and by when?" and then translate that knowledge back into actions to take within one or more planning horizons.

Climate change adaptation includes actions that enable ecosystems and people to better cope with or adjust to changing conditions. Some have categorized major strategies into three areas, including resistance, resilience, and facilitated transformation [47–49]. Where vulnerability assessments indicate low vulnerability over upcoming decades, management can concentrate on resistance-based strategies; aiming to prevent ecosystem degradation. Where moderate to high vulnerability is indicated, strategies focused on restoring resilience are the priority. Where vulnerability is indicated as being very high over upcoming decades, options for facilitated transformation need to be identified.

Results of this analysis suggest adaptation strategies that suit the character of the vegetation type. For example, as described in the results, warm desert shrublands and semi-desert grassland types already score into the high vulnerability range. It would be prudent for planners and managers to evaluate current landscape patterns and identify zones where they can anticipate plant invasions from neighboring vegetation [50]. Where degraded from prior land uses, restoration of native herb diversity and nitrogen fixing taxa are also needed. Monitoring for pollinator population trends, invasive plant expansion, and shrub regeneration, are also increasingly urgent.

Further north and upslope, pinyon-juniper woodlands currently tend to score in the low-moderate range of vulnerability, but my mid-century, they score in the moderate to high range of vulnerability. Actions to maintain or restore resilience in these forests are needed [51]. These could include protection of remaining "old growth" stands while restoring natural wildfire regimes and tree canopy densities in the surroundings. Over upcoming decades, as temperature and precipitation patterns change, models of wildfire regimes will need to be updated and customized to local conditions. Monitoring for invasive plant expansion, such as from cheatgrass (*Bromus tectorum*), effects of drought stress, and tree regeneration will all increase in urgency.

Looking out to the upcoming decades towards the mid-21st century, nearly all types assessed here would benefit from a set of resilience-based strategies, so these investments in the near-term may limit needs for more extreme measures later in the century.

4.2. Methodological Issues

The HCCVI could be considered an Indicator-Based Vulnerability Assessment (IBVA), where a series of indicators for exposure or resilience are measured and then combined to approximate relative climate change vulnerability. Tonmoy et al. [29] completed a meta-analysis of vulnerability assessments and provide useful insights into strengths and weaknesses in methods for IBVA design and indicator aggregation. Our HCCVI could be categorized as a hybrid IBVA that combines simulation modeling (e.g., for project climate conditions) with the present value of other indicators (e.g., for sensitivity and adaptive capacity measures) to arrive at a vulnerability score. We normalized each indicator and used an arithmetic mean for their combination. This approach to aggregation requires a high level of independence among component indicators and considerable knowledge of the relationship between each indicator and overall vulnerability. While these vegetation types are well understood and documented, and indicators used here reflect high-quality data, we still cannot presume all assumptions have been fully met with regards to aggregation rules.

In contrast to some of the relatively few other assessment methods for natural community types, we have limited our scope to well-described types characterized by native vegetation. Where others have chosen to assess socio-ecological systems [17,18], the level of uncertainty increases considerably. While we acknowledge the relative utility of formally integrating human dimensions into ecosystem concepts, we feel that the introduced uncertainty is considerable. Our approach aims to limit this uncertainty by first assessing the natural community type, and then in subsequent steps, these outputs can be brought together with human dimensions affecting a given landscape of interest.

Also, in contrast to many other assessments, here we have emphasized measuring climate exposure along a trend line from the mid-20th century baseline through recent conditions, and into the future time frame of the upcoming decades. This trend-based approach helps to address the considerable uncertainty associated with climate projections by first grounding "current" measures from already observed climate change, and then by focusing on the upcoming decades though the mid-21st century.

One of the most significant methodological challenges to our framework is the application of the climate exposure measure. We anticipate much additional effort to better tie climate trend data more directly to driving ecological processes (e.g., biomass productivity, hydrologic regime, fire regime) to provide more robust predictions. Similarly, some factors affecting resilience will change over upcoming decades, and so the ability to create reliable forecasts of changing conditions, such as those resulting from future development patterns or invasive species spread, will add precision to overall resilience forecasts. Other challenges identified with this framework included the treatment of functional species groups and "keystone" species. While both concepts for vulnerability measures are desirable and likely provide important contributions, limits to current knowledge become apparent when one attempts to identify and assess species for each category. In addition, locating data sets for component measures that span the range of a given type will remain a challenge. A systematic and regional approach such as this that we have taken can highlight needs for investments in data sets critical to addressing climate change.

5. Conclusions

This framework is intended for widespread application and accumulation of results for many vegetation types. The summary spatial units chosen, including 100 km^2 hexagons and ecoregion units, appear to be practical for this purpose. Some of the input raster layers (30–800 m pixel resolution) are also quite useful for subsequent application to decision making.

These outputs are also compatible with efforts to gauge the range-wide conservation status or risk of range-wide ecosystem collapse. For example, under the IUCN Red List of Ecosystems framework [52,53], Criterion C3 addresses environmental degradation over a 50-year timeframe including the current time period where degradation is expressed in terms of relative proportional extent of and ecosystem type affected at varying levels of relative severity. Since our results can express relative severity (i.e., very high–low climate change vulnerability) in 100 km^2 increments across the range-wide extent of the type, they could apply directly to measuring C3 for red listing.

We believe that the framework illustrated provides a practical basis for accumulating spatially explicit scores for climate change vulnerability of major vegetation types. It also includes sufficient specificity to inform adaptive management responses. Continued investment in this type of analysis, encompassing more types and across national and international scales, should yield benefits to natural resource managers and conservation practitioners as they navigate the challenges posed by climate change over the upcoming decades.

Supplementary Materials: The following are available online at https://databasin.org/galleries/6704179ca499490bafd2e9080df1908a.

Author Contributions: Conceptualization, P.J.C. and H.H.H.; Data curation, J.C.H., S.L.A. and R.L.S.; Formal analysis, P.J.C., J.C.H., S.L.A. and M.M.K.; Funding acquisition, P.J.C. and H.H.H.; Investigation, P.J.C., J.C.H., M.S.R., S.L.A. and K.A.S.; Methodology, P.J.C., J.C.H., M.S.R., S.L.A., H.H.H. and M.M.K.; Project administration, P.J.C. and M.S.R.; Software, J.C.H. and S.L.A.; Supervision, H.H.H.; Validation, P.J.C., J.C.H., S.L.A., R.L.S. and M.M.K.; Visualization, P.J.C., M.S.R., S.L.A., H.H.H., R.L.S. and M.M.K.; Writing—original draft, P.J.C.; Writing—review & editing, J.C.H., S.L.A., H.H.H. and M.M.K.

Funding: This research was funded by the USDOI Bureau of Land Management, cooperative agreement number L13AC00243.

Acknowledgments: We wish to acknowledge the support of U.S. Department of Interior agencies, primarily the Bureau of Land Management and U.S. Fish and Wildlife Service, who have provided resources for this research. Numerous experts have provided review and insights throughout the design of this method and its implementation. Geoff Hammerson conducted climate change vulnerability assessments for animal species involved in this research. Patrick McIntyre provided technical interpretation for some vegetation types, and Don Faber-Langendoen provided editorial review. Mary Harkness and Kristin Snow provided essential support for database design and management.

Conflicts of Interest: The authors declare no conflict of interest. The funders had no role in the design of the study; in the collection, analyses, or interpretation of data; in the writing of the manuscript, or in the decision to publish the results.

Appendix A. Methods Detail for Resilience Measures

Individual measures for resilience were applied to each type. This appendix provides detailed explanation of procedures used for each measurement.

Resilience—Ecosystem Sensitivity

Below are described specific measures of sensitivity, including Landscape Condition, Invasive Plant Species, Fire Regime Departure, and Forest Insect and Disease Risk.

Sensitivity—Landscape Condition

Since human land uses, such as built infrastructure for transportation, urban development, industry, agriculture and other vegetation alterations, are depicted in maps that are periodically updated, they can be used in spatial models to make inferences about the status and trends in human-induced stress and ecological condition of ecosystems and landscapes at regional to global scales [54–57]. The spatial model of landscape condition used here [58] built on a growing body of published methods and software tools for ecological effects assessment and spatial modeling; all of which aim to characterize relative ecological condition of landscapes [59–61]. The model uses regionally available spatial data to transparently express user knowledge regarding the relative effects of land uses on natural ecosystems.

Values close to 1.0 indicate almost no measurable ecological impact from the land use at a given pixel. As described in [57], model parameters were calibrated, and subsequently validated using tens of thousands of field observations indicating relative ecological condition. The result is a map surface that provides relative index scores per pixel between 0.0 and 1.0. Calibration of this model against over 50,000 field occurrences ranked as A = excellent, B = good, C = fair, and D = poor condition was used to identify thresholds in the 0.0–1.0 scale for applications. In this instance, we used one standard deviation above the mean of the index value for the D occurrences to determine the C. vs. D threshold. The overall threshold value breaks are as follows; A-Rank $\geq$ 0.36, B-Rank $\geq$ 0.30, C-Rank $\geq$ 0.25, D-Rank < 0.25. Per pixel scores were summarized to average values per vegetation type per 100 km^2 hexagon for display.

Sensitivity—Invasive Plant Species

Among desert shrubland and steppe, the effects of invasive species on ecosystem integrity is well known and there is considerable concern for their interactions with climate change [61]. Spatial models depicting likely presence and abundance of invasive annual grasses provide an important indication of vegetation condition, and therefore, relative sensitivity under the HCCVI framework.

See [23] and [62] for further explanation of spatial models used here. Using the master database of over 20,000 invasive plant locality records with satellite imagery and a suite of environmental variables, inductive modeling was completed using Random Forests [41]. The resultant independently evaluated map surfaces represent invasive annual grass presence in five categories of expected absolute cover (<5%, 5–15%, 16–25%, 26–45%, and >45%). The five models were then combined onto one surface with higher predicted invasive cover classes taking precedence over lower cover classes on a per pixel basis. These absolute cover values were translated to index scores to reflect "1.0 = most favorable" to "0.0 = least favorable" index values as follows: <5% = 1.0, 5–15% = 0.80, 16–25% = 0.6, 26–45% = 0.4, >45% = 0.2. These per pixel scores were then summarized to average values per vegetation type per 100 km^2 hexagon. This measure applied to desert shrubland and grassland vegetation types where invasive annual grasses have substantial impact.

Sensitivity—Fire Regime Departure

Using estimates of fire frequency and successional rates, fire regime models predict the relative proportion of natural successional stages one might expect to encounter for a community type across a given landscape. They are therefore useful for indicating ecosystem degradation due to wildfire suppression or other human-caused alteration [63]. The US Interagency LANDFIRE program provides both quantitative reference models of vegetation states (i.e., successional stages) and transitions, as well as spatial models of wildfire regime departure (measured in percent of departure) that compare observed vs. predicted aerial extent of each successional stage [33]. For each vegetation type treated in this project, these percent departure values (in 10% increments) were translated to index scores to reflect "1.0 = most favorable" to "0.0 = least favorable" index values as follows: FRCC 1 = 1.0, FRCC 2 = 0.5, and FRCC 3 = 0.15. These per pixel scores were then summarized to average values per vegetation type per 100 km^2 hexagon.

Sensitivity—Forest Insect and Disease Risk

Forest insect and disease impacts on Western US forests and woodlands are becoming pronounced, especially with increasing frequency of relatively mild winters [64]. With increasing rates of overwintering survival of both native and introduced insects, as well as compounded effects of drought [65] there is increasing potential for substantial disruption in forest stand structure, composition, and interacting effects with other natural disturbance processes [66]. The National Insect and Disease Risk Map defines forest areas where, *"the expectation that, without remediation, at least 25% of standing live basal area greater than one inches in diameter will die over a 15-year timeframe (2013–2027) due to insects and diseases"* [67]. The resultant 240 m pixel resolution map represents insect and disease risk along a 0.0–1.0 ramp depicting low to high severity of predicted biomass loss (e.g., 0.05 = 5%, 0.25 = 25%, 0.35 = 35%, etc.). These index values were flipped in order to reflect our "1.0 = most favorable" to "0.0 = least favorable" index values. These per pixel scores were then summarized to average values per vegetation type per 100 km^2 hexagon. This index was applied only to forest and woodland types where forest insects and diseases have substantial impact.

Resilience—Ecosystem Adaptive Capacity

Below are described several measures of adaptive capacity, including diversity within functional species groups, climate change vulnerability of "keystone species," and topo-climate variability.

Diversity within Characteristic Functional Species Groups

Natural communities may include several functional groups, or groups of organisms that pollinate, graze, disperse seeds, fix nitrogen, decompose organic matter, depredate smaller organisms, or perform other functions [68,69]. Functional species groups (FSGs) form a link between key ecosystem processes and structures and ecological resilience. Experimental evidence gathered over recent decades supports the theoretical prediction that communities with functional groups made up of increasingly diverse members tend to be more resilient to perturbations [27]. Therefore, the more diverse the FSG (as measured by taxonomic richness), the greater the likelihood that at least one taxon will have characteristics that allow it to continue to perform its function in the community as climate changes.

Approaches to identifying FSGs for natural communities center on analysis of specific traits in response to environmental constraints [70]. In this effort, environmental settings, dynamics processes, species responses to those settings and processes, and key biotic interactions required for maintenance of each vegetation type, were evaluated to identify the most critical ecological processes and their related FSG. FSGs applicable to the vegetation types in this project included nitrogen fixers, biotic pollinators, biotic seed dispersers, biological soil crusts, perennial cool season grasses, perennial warm season graminoids, halophytes, xerophytes, and pyrophytes. Within each identified group, a listing of characteristic species is included based on existing documentation for the type. Lists of functional species groups used are described in the supporting information (Supplementary Materials).

In each instance, available literature was reviewed to document each group, and for each vegetation type, score them along a 0.0 to 1.0 scale. Due to limited knowledge of variation in FSG composition across the Western US, the same score was applied consistently across the entire distribution of each vegetation type, and we scored FSG diversity in three categories of Low, Medium, or High. While ranges varied by FSG, generally those groups with 1–5 species were scored as low (=0.15, 6–15 species as medium (0.5), and >15 species as high (1.0). Where several FSGs were identified for a given vegetation type, the lowest scoring FSG was applied to the type overall, as it would likely have a strong controlling effect on resilience.

CC Vulnerability of Keystone Species

To assess community resilience, it is important to consider the relative climate change vulnerability of species that play particularly important functional roles. We use the term "keystone species" here to refer to any species that, due to their key functional role in the community, if extirpated or reduced in abundance, could cause disproportionate effects on the populations of other species that characterize the community. Determining the species that can be considered keystone requires an understanding of the natural history of many species in the community being assessed. Although there are quantitative means of identifying keystone species via food web analysis [71], these methods can be time and data intensive. However, identification of potential keystone species may follow directly from the above process "diversity within functional species groups". That is, if an important ecosystem function is represented by just one species, that species is likely providing some "keystone" function for the purposed of this analysis. We reviewed all lists of species across our FSGs and identified a set of keystone species for each vegetation type.

We assessed keystone species vulnerability using the NatureServe Climate Change Vulnerability Index (CCVI), a trait-based tool that allows relatively rapid assessment of suites of species and is applicable to all terrestrial and aquatic plant and animal species [24]. The CCVI places species on a categorical scale from extremely vulnerable to those likely to benefit from climate change. For this effort, the CCVI was applied to the distribution of each species within each of the primary ecoregions that make up the distribution of the vegetation type. The CCVI categories were translated to a numerical scale (0.0–1.0 scale) for combination with other adaptive capacity measures.

Topo-Climatic Variability

Natural communities occur across a range of both macro and micro-climates. For example, some major temperate grassland types of the Western Great Plains form the upland 'matrix' of an ecoregion and in topography of limited relief, while other woodland types occur in microclimates formed by rugged canyons and low mountain ranges. Their current distributions are largely based on both regional and local-scale responses to temperature and precipitation. The relative variability in climate encompassed by the distribution of a given community can provide another useful indication of adaptive capacity under changing climate [72]. The idea of climate change 'velocity' [73] has been proposed as a measure of climate change exposure, and it captures in part the interaction of changing climate with topography. The measure is derived by dividing the rate of projected climate change in units of °C per year by the rate of spatial climate variability, i.e., the temperature differential of adjacent grid cells, measured in °C km^{-1}. Areas with relatively rugged topography and elevational gradients will support a greater diversity of microclimate conditions (re: low velocity) as compared with areas of flat topography. Therefore, for the same increment of climate change and time-period, a given species would be required to migrate a shorter distance in areas of rugged topography as compared with expansive flat landscapes.

Since we provide for independent measures of climate change exposure with the HCCVI framework, we used maps of terrain ruggedness to express the influence of topography on microclimate variability. The terrain ruggedness index (TRI) provided by Riley et al. [74] was used with 90 m digital elevation data of North America. TRI is the sum change in elevation between a given grid cell and its eight neighboring grid cells. For example, a cell located at 200 m elevation, surrounded by four cells at 100 m and four more cells at 125 m would yield a TRI of 700 (400 + 500 − 200 = 700). The topo-climatic variability map was derived by normalizing TRI scores to the 0.01–1.0 scale using extreme TRI estimates as projected for North America (TRI = 6196). We overlaid this normalized map with distributions of each vegetation type to arrive at per pixel scores and then summarized them by 100 km^2 hexagon.

Appendix B. Proportional Area for All Component and Composite HCCVI Scores for 52 Major Vegetation Types (*Supplied as Separate File*)

Again, the summarized results for overall climate change vulnerability of the 52 assessed types are found in Appendix B. Type-specific map, text, and tabular summary information is accessible within Supplementary Materials (see: https://databasin.org/galleries/6704179ca499490bafd2e9080df1908a).

References

1. Swetnam, T.W.; Betancourt, J.L. Mesoscale disturbance and ecological response to decadal climatic variability in the American Southwest. *J. Clim.* **1998**, *11*, 3128–3147. [CrossRef]
2. Wells, P.V. Paleobiogeography of montane islands in the Great Basin since the last glaciopluvial. *Ecol. Monogr.* **1983**, *53*, 341–382. [CrossRef]
3. Betancourt, J.L.; Van Devender, T.R.; Martin, P.S. (Eds.) *Packrat Middens: The Last 40,000 Years of Biotic Change*; University of Arizona Press: Tucson, AZ, USA, 1990.
4. Barros, R.V.; Field, C.B.; Dokken, D.J.; Mastrandrea, M.D.; Mach, K.J.; Bilir, T.E.; Chatterjee, M.; Ebi, K.L.; Estrada, Y.O.; Genova, R.C.; et al. *Climate Change 2014: Impacts, Adaptation, and Vulnerability-Part B: Regional Aspects*; Cambridge University Press: New York, NY, USA, 2014.
5. Mantyka-pringle, C.S.; Martin, T.G.; Rhodes, J.R. Interactions between climate and habitat loss effects on biodiversity: A systematic review and meta-analysis. *Glob. Chang. Biol.* **2012**, *18*, 1239–1252. [CrossRef]
6. Allen, C.D.; Birkeland, C.; Stuart Chapin, F., III; Groffman, P.M.; Guntenspergen, G.R.; Knapp, A.; McGuire, A.D.; Mulholland, P.J.; Peters, D.P.C.; Roby, D.D.; et al. *Thresholds of Climate Change in Ecosystems*; A Report by the US Climate Change Science Program and the Subcommittee on Global Change Research; US Geological Survey, Department of the Interior: Washington, DC, USA, 2009.
7. Finch, D.M. *Climate Change in Grasslands, Shrublands, and Deserts of the Interior American West: A Review and Needs Assessment*; US Department of Agriculture, Forest Service, Rocky Mountain Research Station: Fort Collins, CO, USA, 2012.
8. Thomas, C.D.; Cameron, A.; Green, R.E.; Bakkenes, M.; Beaumont, L.J.; Collingham, Y.C.; Erasmus, B.F.N.; De Siqueira, M.F.; Grainger, A.; Hannah, L.; et al. Extinction risk from climate change. *Nature* **2004**, *427*, 145. [CrossRef] [PubMed]
9. Laidre, K.L.; Stirling, I.; Lowry, L.F.; Wiig, Ø.; Heide-Jørgensen, M.P.; Ferguson, S.H. Quantifying the sensitivity of Arctic marine mammals to climate-induced habitat change. *Ecol. Appl.* **2008**, *18*, S97–S125. [CrossRef] [PubMed]
10. Rowland, E.L.; Davison, J.E.; Graumlich, L.J. Approaches to evaluating climate change impacts on species: A guide to initiating the adaptation planning process. *Environ. Manag.* **2008**, *47*, 322–337. [CrossRef] [PubMed]

11. Pacifici, M.; Foden, W.B.; Visconti, P.; Watson, J.E.M.; Butchart, S.H.; Kovacs, K.M.; Scheffers, B.R.; Hole, D.G.; Martin, T.G.; Akçakaya, H.R.; et al. Assessing species vulnerability to climate change. *Nat. Clim. Chang.* **2015**, *5*, 215. [CrossRef]

12. Foden, W.B.; Young, B.E.; Akçakaya, H.R.; Garcia, R.A.; Hoffmann, A.A.; Stein, B.A.; Thomas, C.D.; Wheatley, C.J.; Bickford, D.; Carr, J.A.; et al. Climate change vulnerability assessment of species. *Wiley Interdiscip. Rev. Clim. Chang.* **2018**, *10*, e551. [CrossRef]

13. Swanston, C.; Janowiak, M.; Iverson, L.; Parker, L.; Mladenoff, D.; Brandt, L.; Butler, P.; Pierre, M.S.; Prasad, A.; Matthews, S.; et al. *Ecosystem Vulnerability Assessment and Synthesis*; US Department of Agriculture, Forest Service, Rocky Mountain Research Station: Fort Collins, CO, USA, 2010.

14. Rustad, L.; Campbell, J.; Dukes, J.S.; Huntington, T.; Lambert, K.F.; Mohan, J.; Rodenhouse, N. *Changing Climate, Changing Forests: The Impacts of Climate Change on Forests of the Northeastern United States and Eastern Canada*; US Department of Agriculture, Forest Service, Rocky Mountain Research Station: Fort Collins, CO, USA, 2012.

15. Schneider, R.R. *Alberta's Natural Subregions under a Changing Climate: Past, Present, and Future*; Alberta Biodiversity Monitoring Institute: Edmonton, AB, Canada, 2013.

16. Hamann, A.; Roberts, D.R.; Barber, Q.E.; Carroll, C.; Nielsen, S.E. Velocity of climate change algorithms for guiding conservation and management. *Glob. Chang. Biol.* **2015**, *21*, 997–1004. [CrossRef]

17. Lindner, M.; Maroschek, M.; Netherer, S.; Kremer, A.; Barbati, A.; Garcia-Gonzalo, J.; Seidl, R.; Delzon, S.; Corona, P.; Kolström, M.; et al. Climate change impacts, adaptive capacity, and vulnerability of European forest ecosystems. *For. Ecol. Manag.* **2010**, *259*, 698–709. [CrossRef]

18. Cinner, J.E.; Huchery, C.; Darling, E.S.; Humphries, A.T.; Graham, N.A.J.; Hicks, C.C.; Marshall, N.; McClanahan, T.R. Evaluating Social and Ecological Vulnerability of Coral Reef Fisheries to Climate Change. *PLoS ONE* **2013**, *8*, e74321. [CrossRef] [PubMed]

19. Thorne, J.H.; Boynton, R.M.; Holguin, A.J.; Stewart, J.A.; Bjorkman, J. *A Climate Change Vulnerability Assessment of California's Terrestrial Vegetation*; California Department of Fish and Wildlife: Sacramento, CA, USA, 2016.

20. Flint, L.E.; Flint, A.L.; Thorne, J.H.; Boynton, R.M. Fine-scale hydrological modeling for regional applications: Model development and performance. *Ecol. Process.* **2013**, *2*, 25. [CrossRef]

21. Li, D.; Wu, S.; Liu, L.; Zhang, Y.; Li, S. Vulnerability of the global terrestrial ecosystems to climate change. *Glob. Chang. Boil.* **2018**, *24*, 4095–4106. [CrossRef] [PubMed]

22. Comer, P.J.; Young, B.; Schulz, K.; Kittel, G.; Unnasch, B.; Braun, D.; Hammerson, G.; Smart, L.; Hamilton, H.; Auer, S.; et al. *Climate Change Vulnerability and Adaptation Strategies for Natural Communities: Piloting Methods in the Mojave and Sonoran Deserts*; Report to US Fish and Wildlife Service; NatureServe: Arlington, VA, USA, 2012.

23. Comer, P.J.; Crist, P.J.; Reid, M.S.; Hak, J.; Hamilton, H.; Unnasch, B.; Kutner, L. *A Rapid Ecoregional Assessment of the Central Basin and Range Ecoregion. Report, Appendices, and Databases Provided to the Bureau of Land Management.* Available online: https://landscape.blm.gov/geoportal/catalog/REAs/REAs.page (accessed on 5 July 2019).

24. Young, B.; Byers, E.; Gravuer, K.; Hall, K.; Hammerson, G.; Redder, A. *Guidelines for Using the NatureServe Climate Change Vulnerability Index*; NatureServe: Arlington, VA, USA, 2011.

25. Holling, C.S. Resilience and stability of ecological systems. *Annu. Rev. Ecol. Syst.* **1973**, *4*, 1–23. [CrossRef]

26. Gunderson, L.H. Ecological resilience—In theory and application. *Annu. Rev. Ecol. Syst.* **2000**, *31*, 425–439. [CrossRef]

27. Walker, B.; Holling, C.S.; Carpenter, S.R.; Kinzig, A. Resilience, adaptability and transformability in social–ecological systems. *Ecol. Soc.* **2004**, *9*. [CrossRef]

28. Magness, D.R.; Morton, J.M.; Huettmann, F.; Chapin III, F.S.; McGuire, A.D. A climate-change adaptation framework to reduce continental-scale vulnerability across conservation reserves. *Ecosphere* **2011**, *2*, 1–23. [CrossRef]

29. Tonmoy, F.N.; El-Zein, A.; Hinkel, J. Assessment of vulnerability to climate change using indicators: A meta-analysis of the literature. *Wiley Interdiscip. Rev. Clim. Chang.* **2014**, *5*, 775–792. [CrossRef]

30. Wiken, E.; Jiménez Nava, F.; Griffith, G. *North American Terrestrial Ecoregions—Level III*; Commission for Environmental Cooperation: Montreal, QC, Canada, 2011.

31. Comer, P.; Faber-Langendoen, D.; Evans, R.; Gawler, S.; Josse, C.; Kittel, G.; Snow, K. *Ecological Systems of the United States: A Working Classification of US Terrestrial Systems*; NatureServe: Arlington, VA, USA, 2003.

32. Comer, P.J.; Schulz, K.A. Standardized ecological classification for mesoscale mapping in the southwestern United States. *Rangel. Ecol. Manag.* **2007**, *60*, 324–335. [CrossRef]

33. Rollins, M.G. LANDFIRE: A nationally consistent vegetation, wildland fire, and fuel assessment. *Int. J. Wildland Fire* **2009**, *18*, 235–249. [CrossRef]

34. Comer, P.J.; Hak, J.C.; Josse, C.; Smyth, R. Long-term Change in Extent and Current Protection of Terrestrial Ecosystem Diversity in the Temperate and Tropical Americas. *PLoS ONE* **2019**. Unpublished work.

35. Kling, M.M.; Auer, S.; Comer, P.J.; Ackerly, D.D.; Hamilton, H.H. Multiple dimensions of ecological vulnerability to climate change. *Global Change Biology* **2019**. Unpublished work.

36. O'Donnell, M.S.; Ignizio, D.A. Bioclimatic predictors for supporting ecological applications in the conterminous United States. *US Geol. Surv. Data Ser.* **2012**, *691*, 1–10.

37. Sheehan, T.; Bachelet, D. and Ferschweiler, K.; Projected major fire and vegetation changes in the Pacific Northwest of the conterminous United States under selected CMIP5 climate futures. *Ecol. Model.* **2015**, *317*, 16–29. [CrossRef]

38. Oyler, J.W.; Ballantyne, A.; Jencso, K.; Sweet, M.; Running, S.W. Creating a topoclimatic daily air temperature dataset for the conterminous United States using homogenized station data and remotely sensed land skin temperature. *Int. J. Climatol.* **2015**, *35*, 2258–2279. [CrossRef]

39. Daly, C.; Halbleib, M.; Smith, J.I.; Gibson, W.P.; Doggett, M.K.; Taylor, G.H.; Curtis, J.; Pasteris, P.P. Physiographically sensitive mapping of climatological temperature and precipitation across the conterminous United States. *Int. J. Clim.* **2008**, *28*, 2031–2064. [CrossRef]

40. Wang, T.; Hamann, A.; Spittlehouse, D.; Carroll, C. Locally downscaled and spatially customizable climate data for historical and future periods for North America. *PLoS ONE* **2016**, *11*, e0156720. [CrossRef]

41. Liaw, A.; Wiener, M. Classification and regression by randomForest. *R News* **2002**, *2*, 18–22.

42. Bahn, V.; McGill, B.J. Testing the predictive performance of distribution models. *Oikos* **2013**, *122*, 321–331. [CrossRef]

43. Gregorutti, B.; Michel, B.; Saint-Pierre, P. Grouped variable importance with random forests and application to multiple functional data analysis. *Comput. Stat. Data Anal.* **2015**, *90*, 15–35. [CrossRef]

44. Damschen, E.I.; Baker, D.V.; Bohrer, G.; Nathan, R.; Orrock, J.L.; Turner, J.R.; Brudvig, L.A.; Haddad, N.M.; Levey, D.J.; Tewksbury, J.J. How fragmentation and corridors affect wind dynamics and seed dispersal in open habitats. *Proc. Natl. Acad. Sci. USA* **2014**, *111*, 3484–3489. [CrossRef] [PubMed]

45. Pausas, J.G.; Keeley, J.E. Abrupt climate-independent fire regime changes. *Ecosystems* **2014**, *17*, 1109–1120. [CrossRef]

46. Aycrigg, J.L.; Davidson, A.; Svancara, L.K.; Gergely, K.J.; McKerrow, A.; Scott, J.M. Representation of ecological systems within the protected areas network of the continental United States. *PLoS ONE* **2013**, *8*, e54689. [CrossRef] [PubMed]

47. Keith, D.A.; Rodriguez, J.P.; Rodríguez-Clark, K.M.; Nicholson, E.; Aapala, K.; Alonso, A.; Asmüssen, M.; Bachman, S.; Basset, A.; Barrow, E.G.; et al. Scientific Foundations for an IUCN Red List of Ecosystems. *PLoS ONE* **2013**, *8*, e62111. [CrossRef] [PubMed]

48. Bland, L.; Keith, D.A.; Miller, R.M.; Murray, N.J.; Rodríguez, J.P. *Guidelines for the Application of IUCN Red List of Ecosystems Categories and Criteria, Version 1.1*; IUCN: Gland, Switzerland, 2016.

49. Hansen, L.; Biringer, J.L.; Hoffman, J. *Buying Time: A User's Manual to Building Resistance and Resilience to Climate Change in Natural Systems*; WWF: Gland, Switzerland, 2003.

50. Millar, C.I.; Stephenson, N.L.; Stephens, S.L. Climate change and forests of the future: Managing in the face of uncertainty. *Ecol. Appl.* **2007**, *17*, 2145–2151. [CrossRef] [PubMed]

51. Chambers, J.C.; Beck, J.L.; Campbell, S.; Carlson, J.; Christiansen, T.J.; Clause, K.J.; Dinkins, J.B.; Doherty, K.E.; Griffin, K.A.; Havlina, D.W.; et al. *Using Resilience and Resistance Concepts to Manage Threats to Sagebrush Ecosystems, Gunnison Sage-Grouse, and Greater Sage-Grouse in Their Eastern Range: A Strategic Multi-Scale Approach*; Department of Agriculture, Forest Service, Rocky Mountain Research Station: Fort Collins, CO, USA.

52. Bachelet, D.; Ferschweiler, K.; Sheehan, T.; Strittholt, J. Climate change effects on southern California deserts. *J. Arid Environ.* **2016**, *127*, 17–29. [CrossRef]

53. Thorne, J.H.; Choe, H.; Stine, P.A.; Chambers, J.C.; Holguin, A.; Kerr, A.C.; Schwartz, M.W. Climate change vulnerability assessment of forests in the Southwest USA. *Clim. Chang.* **2018**, *148*, 387–402. [CrossRef]

54. Sanderson, E.W.; Jaiteh, M.; Levy, M.A.; Redford, K.H.; Wannebo, A.V.; Woolmer, G. The human footprint and the last of the wild: The human footprint is a global map of human influence on the land surface, which suggests that human beings are stewards of nature, whether we like it or not. *AIBS Bull.* **2002**, *52*, 891–904.
55. Theobald, D.M. A general model to quantify ecological integrity for landscape assessments and US application. *Landsc. Ecol.* **2013**, *28*, 1859–1874. [CrossRef]
56. Haddad, N.M.; Brudvig, L.A.; Clobert, J.; Davies, K.F.; Gonzalez, A.; Holt, R.D.; Lovejoy, T.E.; Sexton, J.O.; Austin, M.P.; Collins, C.D.; et al. Habitat fragmentation and its lasting impact on Earth's ecosystems. *Sci. Adv.* **2015**, *1*, e1500052. [CrossRef]
57. Hak, J.C.; Comer, P.J. Modeling landscape condition for biodiversity assessment—application in temperate North America. *Ecol. Indic.* **2017**, *82*, 206–216. [CrossRef]
58. Riitters, K.H.; Wickham, J.D. How far to the nearest road? *Front. Ecol. Environ.* **2003**, *1*, 125–129. [CrossRef]
59. Hansen, A.J.; Knight, R.L.; Marzluff, J.M.; Powell, S.; Brown, K.; Gude, P.H.; Jones, K. Effects of exurban development on biodiversity: Patterns, mechanisms, and research needs. *Ecol. Appl.* **2005**, *15*, 1893–1905. [CrossRef]
60. Leu, M.; Hanser, S.E.; Knick, S.T. The human footprint in the west: A large-scale analysis of anthropogenic impacts. *Ecol. Appl.* **2008**, *18*, 1119–1139. [CrossRef] [PubMed]
61. Abatzoglou, J.T.; Kolden, C.A. Climate change in western US deserts: Potential for increased wildfire and invasive annual grasses. *Rangel. Ecol. Manag.* **2011**, *64*, 471–478. [CrossRef]
62. Hak, J.C.; Comer, P.J. Modeling Invasive Annual Grass Vulnerability in the Cold Deserts of the Intermountain West. *Rangel. Ecol. Manag.* **2019**. Unpublished work.
63. Swaty, R.; Blankenship, K.; Hagen, S.; Fargione, J.; Smith, J.; Patton, J. Accounting for ecosystem alteration doubles estimates of conservation risk in the conterminous United States. *PLoS ONE* **2011**, *6*, e23002. [CrossRef] [PubMed]
64. Kurz, W.A.; Dymond, C.C.; Stinson, G.; Rampley, G.J.; Neilson, E.T.; Carroll, A.L.; Ebata, T.; Safranyik, L.; Kurz, W. Mountain pine beetle and forest carbon feedback to climate change. *Nature* **2008**, *452*, 987–990. [CrossRef]
65. Breshears, D.D.; Cobb, N.S.; Rich, P.M.; Price, K.P.; Allen, C.D.; Balice, R.G.; Romme, W.H.; Kastens, J.H.; Floyd, M.L.; Belnap, J.; et al. Regional vegetation die-off in response to global-change-type drought. *Proc. Natl. Acad. Sci. USA* **2005**, *102*, 15144–15148. [CrossRef]
66. Allen, C.D.; Macalady, A.K.; Chenchouni, H.; Bachelet, D.; McDowell, N.; Vennetier, M.; Kitzberger, T.; Rigling, A.; Breshears, D.D.; Hogg, E.; et al. A global overview of drought and heat-induced tree mortality reveals emerging climate change risks for forests. *For. Ecol. Manag.* **2010**, *259*, 660–684. [CrossRef]
67. Krist, F.J., Jr.; Ellenwood, J.R.; Wood, M.E.; McMahan, A.J.; Cowardin, J.P.; Ryerson, D.E.; Sapio, F.J.; Zweifler, M.O.; Romero, A.S. *2027 National Insect and Disease Forest Risk Assessment 2013*; USDA Forest Service, Forest Health Technology Enterprise Team: Fort Collins, CO, USA, 2013.
68. Rosenfeld, J.S. Functional redundancy in ecology and conservation. *Oikos* **2002**, *98*, 156–162. [CrossRef]
69. Folke, C.; Carpenter, S.; Walker, B.; Scheffer, M.; Elmqvist, T.; Gunderson, L.; Holling, C.S. Regime shifts, resilience, and biodiversity in ecosystem management. *Annu. Rev. Ecol. Evol. Syst.* **2004**, *35*, 557–581. [CrossRef]
70. Díaz, S.; Cabido, M. Vive la difference: Plant functional diversity matters to ecosystem processes. *Trends Ecol. Evol.* **2001**, *16*, 646–655. [CrossRef]
71. Ebenman, B.; Jonsson, T. Using community viability analysis to identify fragile systems and keystone species. *Trends Ecol. Evol.* **2005**, *20*, 568–575. [CrossRef] [PubMed]
72. Ackerly, D.D.; Loarie, S.R.; Cornwell, W.K.; Weiss, S.B.; Hamilton, H.; Branciforte, R.; Kraft, N.J.B. The geography of climate change: Implications for conservation biogeography. *Divers. Distrib.* **2010**, *16*, 476–487. [CrossRef]
73. Loarie, S.R.; Duffy, P.B.; Hamilton, H.; Asner, G.P.; Field, C.B.; Ackerly, D.D. The velocity of climate change. *Nature* **2009**, *462*, 1052. [CrossRef] [PubMed]
74. Riley, S.J. Index that quantifies topographic heterogeneity. *Intermt. J. Sci.* **1999**, *5*, 23–27.

 *land*

Article

Assessing the Role of the Perceived Impact of Climate Change on National Adaptation Policy: The Case of Rice Farming in Indonesia

Mohammad Rondhi [1,*], Ahmad Fatikhul Khasan [1,*], Yasuhiro Mori [2] and Takumi Kondo [2]

[1] Department of Agribusiness, University of Jember, Jember 68121, Indonesia
[2] Research Faculty of Agriculture, Hokkaido University, Sapporo 060-8589, Japan;
ymouri@agecon.agr.hokudai.ac.jp (Y.M.); kondot@agecon.agr.hokudai.ac.jp (T.K.)
* Correspondence: rondhi.faperta@unej.ac.id (M.R.); ahmad.fatih@unej.ac.id (A.F.K.)

Received: 12 March 2019; Accepted: 7 May 2019; Published: 10 May 2019

Abstract: Climate change (CC) is one of the primary threats to the agricultural sector in developing countries. Several empirical studies have shown that the implementation of adaptation practices can reduce the adverse effects of CC. The likelihood of farmers performing adaptation practices is mostly influenced by the degree of CC impact that they perceive. Thus, we identified the characteristics of farmers that affect the degree of the CC impact that they perceive. We used data from the Indonesian Rice Farm Household survey consisting of 87,330 farmers. An ordered probit regression model was used to estimate the effect of each variable on the degree of the perceived impact of CC. The results of this study confirm those of previous empirical studies. Several variables that have been identified as having a positive effect on farmer adaptation practices, such as farmer education, land tenure, irrigation infrastructure, cropping system, chemical fertilizer application, access to extension services, and participation in farmer groups, negatively affect the degree of the perceived impact of CC. However, a different result was found in the estimation of the gender variable. We found that female farmers have a higher CC resilience and ability to withstand climatic shocks and risks than male farmers. Female farmers have a more positive perception of future farming conditions than male farmers. We recommend the implementation of a national adaptation policy that use and expand the channel of agricultural extension services to deliver the planned adaptation policy, and prioritizes farmers with insecure land tenure. Additionally, we encourage the increasing of female involvement in the CC adaptation practices and decision-making processes.

Keywords: climate change; perceived impact of climate change; climate change adaptation; ordered probit regression; determinants of climate change impact

---

1. Introduction

Climate change (CC) is a global phenomenon that is harming climate-dependent activity such as agricultural production. The negative impacts of CC on agriculture, both for crop and animal production, have been well documented [1–4]. The degree of adverse impacts of CC on farmers is determined by the vulnerability of those farmers to CC. Vulnerability is the propensity or predisposition of a natural or human system to be adversely affected by CC, and encompasses a variety of concepts and elements, including sensitivity or susceptibility and lack of capacity to cope and adapt [5]. Vulnerability, then, is a function of three aspects: Exposure to hazard, sensitivity to damage, and ability to cope. Currently, there are no means to control the occurrence of natural hazards; increasing the system's adaptive capacity can reduce its sensitivity to damage caused by natural hazards [6]. Based on that definition, the implementation of an appropriate adaptation strategy can minimize a farmer's vulnerability.

An appropriate adaptation strategy is required to moderate the adverse effects of CC [7]. There are two types of adaptation: Autonomous and planned adaptation [8]. In the former, farmers independently adapt their farming practices to the observed climatic change. In the latter, the government plans and implements an adaptation policy. Several studies have reported that farmers in developing countries have adjusted their farming practices in response to CC and found that the adaptation has a positive effect on crop yield [9,10]. However, several barriers limit adaptation practices, such as financial barriers (lack of financial resources and/or lack of supporting institutions, whether public or private, to finance adaptation), social and cultural barriers (individuals and group perspectives, values, and beliefs toward CC), and informational and cognitive barriers (individual perceptions, values, and opinions about the risk of CC) [11]. This study focuses on the third barrier, and specifically individual climate risk perception.

A farmer's perception of climate risk is essential because it represents the degree of perceived impact (P-I)—a measure of how a farmer personally feels about the impact of a particular occurrence [12,13]. Past exposure to climate-related disaster increases the degree of P-I, which in turn drives farmers to undertake adaptation actions [14,15]. While some studies stressed the benefits of autonomous adaptation, other studies reported that it ultimately results in unintended maladaptive outcomes, such as increasing the farmer's vulnerability to CC, shifting the vulnerability to other stakeholders or sectors, and decreasing the quality of common pooled resources [16–19]. Thus, assessing a farmer's P-I toward CC is essential in two aspects: First, it provides valuable information about the efforts to encourage autonomous adaptation; second, it provides crucial insight into the effort to avoid maladaptation practices. As most developing countries have a national adaptation policy [20], this study contributes to addressing the question of which farmers should be prioritized and through what channel the content of a policy should be delivered. Figure 1 shows how climate risk perception is related to autonomous adaptation and adaptation outcome.

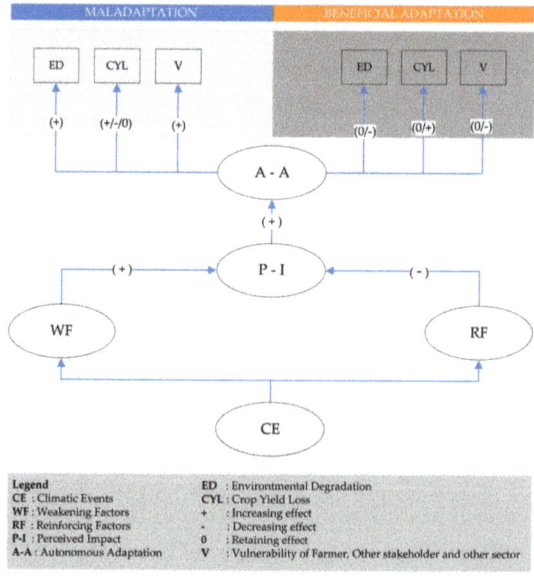

Figure 1. The relationship between climate change risk perception, autonomous adaptation, and adaptation outcome.

While CC is a global issue that impacts farmers in both developed and developing countries, it causes more severe damage to those in developing countries [12,21], since the majority of these farmers are poor and less adaptable [22]. These farmers are often located in less favorable agricultural areas, which increases their vulnerability [23]. The availability of adaptation instruments, whether of a physical nature, such as stress-tolerant crops, or of an institutional nature, such as crop insurance, is limited compared to those in developed countries [24]. Farmers, especially in developing countries, treat CC as a threat to their farming only when they perceive that it significantly decreases crop yield [25,26]. For example, the farmer who experiences crop yield loss due to CC may feel that the impact is not significant because the loss is relatively small compared to their wealth, or due to other reasons, and it is less likely that they will adopt any adaptation strategy. Conversely, a farmer who perceives that the impact of CC is high is more likely to adjust their farming practices to minimize the effect. Based on this reasoning, a farmer's P-I of CC influences their likelihood of adopting an adaptation strategy. The higher the P-I, the more likely the farmer is to adapt, and vice versa. Since the degree of P-I is affected by many factors, including technical, social, economic, or institutional factors [27–29], identifying factors influencing a farmer's P-I of CC is essential. This information will be useful to determine which farmers should be prioritized and targeted.

Extensive studies on the effect of CC on agriculture in developing countries have been conducted. While CC affects almost all aspects of a developing country's economy, agriculture is the most substantially impacted [30,31]. However, few studies have analyzed the effects of CC using nationally representative farm data [32–34]. Nationally representative studies on the impact of CC on agriculture in developing countries often use aggregate data [21,35–38]; other times, individual farm data are employed, and the study is limited to local and community levels [9,39,40]. Whereas the former is useful in providing an overall view of the CC effect and the latter is useful in informing a detailed picture of how a farmer is affected and manages to adapt, it is necessary to conduct a nationally representative study on how farmers are affected by CC. Most of these studies assessed the actual impact of CC on agriculture, which can be different from the farmer perception.

The role of climate risk perception in CC adaptation has received considerable attention. A study in Bangladesh showed that farmer perceptions of CC are mostly aligned with observed meteorological data and are correlated positively with the rate of adopted adaptation practices [15]. Similarly, a study in French coastal populations showed that they perceive the local changes in climate, weather, coral, and beaches, but they only regard it as a problem instead of a danger [41]. In contrast to the result from France, a study on the peri-urban community in Mexico that experiences a risk of drought indicated that the community perceives CC and treats it as a threat because their livelihood as brick producers is severely impacted by climate change [42]. A study on Canadian bivalve aquaculture indicated the importance of stakeholder perceptions of CC in adapting to these changes and further expanding the industry [43]. A cross-country analysis in Europe indicated that the perception of CC is affected by individual-level factors such as gender, age, political orientation, and education, but the size of the effects of each variable varies across countries [44].

The general purpose of this study was to identify the determinants of farmers' P-I of CC. Specifically, we aimed to determine what factors, economic, social, technical, or institutional, affect whether farmers perceive CC as a threat to their farming. The contribution of this study is two-fold. First, this study provides nationally representative information on the farm-level impact of CC in developing countries. While numerous efforts to study the effect of CC on developing countries have been made, few studies have used nationally representative data. Second, we used a P-I measure of CC (whether a farmer thinks they are not, less, or highly affected by CC), which was found to significantly drive adaptation practices. This information is relevant to Indonesia, where a national adaptation policy is being implemented. This study contributes to the growing discussion of climate risk perception.

2. Background: Climate Change and Its Impact on Indonesian Agriculture

Indonesia is the largest archipelagic country in the world with over 17,000 islands, with a total land area of 190 million ha. Indonesia had 37 million ha of agricultural land in 2018, and rice fields accounted for 8 million ha, of which 58.13% had irrigation infrastructure [45]. The rice fields are spread out over the major island in Indonesia, but most of them (3.1 million ha) are located on the island of Java [45]. Rice is a vital food crop in Indonesia along with maize, soybean, green bean, cassava, and sweet potato. The annual production of rice in Indonesia is 81 million tons of dry unhusked rice. Beside food crops, Indonesian agriculture includes horticulture: Vegetables (shallot, chili, cayenne, garlic, and potato), fruits (mango, banana, citrus, durian, pomelo, and mangosteen), and flowers (chrysanthemum, rose, orchid, and tuberose). Additionally, Indonesia produces plantation crops, medical crops, and livestock.

As in many developing countries, agriculture plays a vital role in the Indonesian economy. Its production accounts for 13% of the Indonesian gross domestic product (GDP) [46], and it provides a livelihood for 25 million farm households [45]. Among various crops cultivated in Indonesia, rice is the primary food crop. Of the 25 million farm households, 17 million are rice farmers with an average land possession of 0.6 ha [47]. As a smallholder farmer, the rice farmer is economically more vulnerable to external shocks, such as those due to climatic change. In Indonesia, rice farming has been severely impacted by climate change. The changing rainfall frequency and intensity, the increase in temperature, and the rise in sea level have significantly contributed to declining rice productivity. The frequency of occurrence of extreme events, such as flood and drought, is increasing, which causes crop loss. Increasing temperature causes the proliferation of pests and diseases [48]. Thus, attempts to mitigate and adapt to the risk of CC is required for the resilience of rice farming in Indonesia.

Resilience toward CC is defined as the ability to withstand climatic shocks and risks [49]. Forsyth [50] provides a comprehensive review of the definition of CC resilience. The early definition stated that CC resilience is related mainly to the physical properties, such as infrastructure and ecosystems, and its stability during an occurrence of shocks. The definition is then improved to include not only physical properties but also socio-economic factors such as diverse access to sources of livelihood. Finally, the definition of resilience is brought into a broader context of wider social processes and transformation. For the sake of clarity, this paper defines a farmer's resilience toward CC as their ability to withstand and minimize the adverse impact of CC on their farming.

The study of CC in Indonesia has been extensive and covers a wide range of aspects such as agriculture [48,51–55], natural disasters and management of coastal areas [56–58], the politics of climate change [59–61], and the public perspective on climate change [62,63]. These studies stated that climate change will affect many aspects of the Indonesian economy, and agriculture will be the hardest hit. In response to this, the government created a National Action Plan (NAP) to mitigate the risk of and to adapt to climate change. One of the primary targets of the NAP is to achieve economic resilience by achieving food security. The primary objective in achieving food security is to reduce production loss due to extreme climatic events and CC [64]. The primary strategy to reduce farm production loss is applying a CC-resilient farming system and CC-adaptive farm technology. The government should ensure that the farmer adopts both of these strategy. However, limited government resources and the large number of farmers will limit the adoption rate of this strategy. Thus, it is essential to identify farmers with a high probability to change and adapt, and to target the implementation of the NAP to these farmers.

3. Methods

3.1. Variable Descriptions

3.1.1. Dependent Variable

The dependent variable is the P-I of climate change on farm yield. A farmer's perception of CC is a subjective measure that represents what impact they think CC has on their farm. Numerous studies have stressed the importance of a farmer's perception toward CC [13,65,66]. This variable was recorded on an ordinal scale to represent the degree of yield loss caused by CC. "No impact" (the farmer perceives no impact of CC on yield), "low impact" (the farmer believes that <50% of yield loss is due to CC), or "high impact" (the farmer believes that <50% of yield loss is due to CC) were possible variables.

3.1.2. Independent Variables

Many factors, including social, economic, technical, institutional, or climatic, may affect the degree of impact of CC on farm production. One factor might reduce the severity of effects because its existence causes farm production to have a stronger resistance toward changing climate so that the farmer can sustain the change without suffering significant yield loss. Alternatively, a factor might reduce the severity of impact because it drives the farmer to take action to limit CC-caused damage. This section briefly reviews the findings of the empirical literature on factors affecting the impact of CC.

The Social Factors

We define social factors as the personal characteristics of a farmer. These factors include the age, education, and gender of the farm household head. Several studies have included these factors in identifying how a farmer perceives and adapts to the impact of CC. Several studies have stated that a farmer's age is correlated positively with their resilience against CC. Older farmers have a greater awareness of the impact of CC [67], and their vast farming experience enables them to implement less costly adaptation methods while sustaining a relatively high level of farm productivity [9]. Similarly, the educated farmer copes with CC better than the less educated one, since they can access better information about CC and adaptation technology [9,65,67]. Finally, the issue of gender in CC has received considerable attention because female farmers are more vulnerable to CC, but they have limited access to resources that can be used to adapt [68,69]. Female farmers are less likely to adopt soil conservation methods, cultivate more diverse crops, or plant trees to reduce the effects of CC [65].

The Economic Factors

The economic factors are asset-related. We include three variables in this group: Land tenure, landholding, and the source of farm capital. Previous studies have shown that farmers with higher wealth tend to better adapt to CC. Land tenure security is a critical factor for CC adaptation, since it encourages farmers to exert more effort and investment in adaptation practices [70,71]. However, a larger land size increases the cost of adaptation and reduces adaptation practices. Previous studies have stated that farmers with access to credit institutions have a higher probability of adapting to CC [9,65,67]. However, having access to credit institutions does not necessarily mean that a farmer will use borrowed money to obtain a high farm budget. Thus, we use farm capital source instead of mere access to credit institutions.

The Technical Factors

We define this category as the technical characteristics of rice farming, which include four variables: Irrigation infrastructure, cropping systems, fertilizer applications, and annual cultivation frequency. Irrigation infrastructure in particular, and agricultural water management in general, play a vital role in mitigating the risk of CC [72–74]. The changing climate alters the frequency of rainfall and affects water availability and crop requirements. Adequate irrigation infrastructure is crucial for the effective distribution of water resources. The farming of mixed species rather than monoculture farming can mitigate the adverse effects of CC [75]. Mixed-species cropping between crops with complementary traits will, with proper management, produce biodiversity and economic advantages in the form of increased productivity. Another significant factor in the technical aspect of farming with respect to CC is fertilizer application [76]. Fertilizer is a primary farm input. However, the excessive use of chemical fertilizer increases the amount of greenhouse gas emissions, which exacerbates CC [77]. The primary challenge of limiting the excessive use of chemical fertilizer is the farmer's perception. A farmer believes that fertilizer application is correlated positively with farm yield. Thus, it is essential to identify how fertilizer application affects how a farmer regards the impact of CC. Similar to the previous variable, the amount of annual rice cultivation increases the amount of fertilizer usage. Additionally, as the amount of cultivation increases, a farmer will be more exposed to the risk of being impacted by CC. It is essential, then, to identify whether a farmer who cultivates rice more frequently perceives a more severe impact.

The Institutional Factors

Several studies have shown the importance of institutional factors in reducing the impact of CC and encouraging farmers to perform adaptation practices. We include three variables in this category: Participation in farmer groups, access to extension services, and participation in farmer field schools. Conceptually, a farmer group is an important tool for the government to distribute and deliver agricultural policy content to farmers. Participation in a farmer group increases the productivity of the farmer [78] and facilitates members to obtain farm input such as fertilizer and seed [79]. Thus, participation in a farmer group has the potential to increase a farmer's resiliency against climate change. To deliver new information and technology, the government specifically established extension services, and access to these services increases farm performance [80]. In the context of climate change, extension services are the leading channel for the provision of information about climate change and adaptation strategies and technology for farmers. Several studies have indicated that access to extension services increases farmer awareness of CC and their adaptation practices [9,65,67]. Similar to the previous institutions, a farmer field school (FFS) is a government-established service that facilitates the dissemination of new knowledge and skills to farmers. A longitudinal study in East Africa stated that participation in an FFS increases farm productivity by 61% and plays a critical role in reducing poverty [81].

The Climatic Factors

Climate change alters the frequency and intensity of rainfall. In some areas, the intensity of rainfall increases, which causes floods, whereas in mountainous areas, increasing rainfall intensity causes landslides. In other areas, the intensity of rainfall decreases, which causes droughts. All of this CC-caused disaster has a substantial impact on agriculture. Floods and landslides cause severe economic damage including loss of crops, whereas drought reduces the amount of harvested farmland and reduces yield [9,82]. Thus, in this category, we include four types of CC-caused disaster—flood, drought, heavy rain, and other hazards (e.g., landslides)—to determine which disaster the farmer perceives as having the most severe impact on their farming.

3.2. Data

We used a nationwide rice farming survey in Indonesia administered by the Indonesian Bureau of Statistics (BPS). The survey was conducted from May 2014 through June 2016 and covered a sample of 87,330 rice farm households (RFHs). The sample selection involved two-stage stratified random sampling. In the first stage, the national BPS office randomly selected census blocks from a total of 844,946 blocks and obtained 8933 sample blocks. Only census blocks with 10 or more RFHs were considered eligible for sample selection. In the second stage, the district BPS office stratified RFHs by land size. Only RFHs with a land size no less than 550 m^2 for lowland rice and 100 m^2 for upland rice were eligible for sample selection. Figure 2 shows, for each province in Indonesia, the distribution of sample RFHs and the percentage of farmers who experienced the effects of CC.

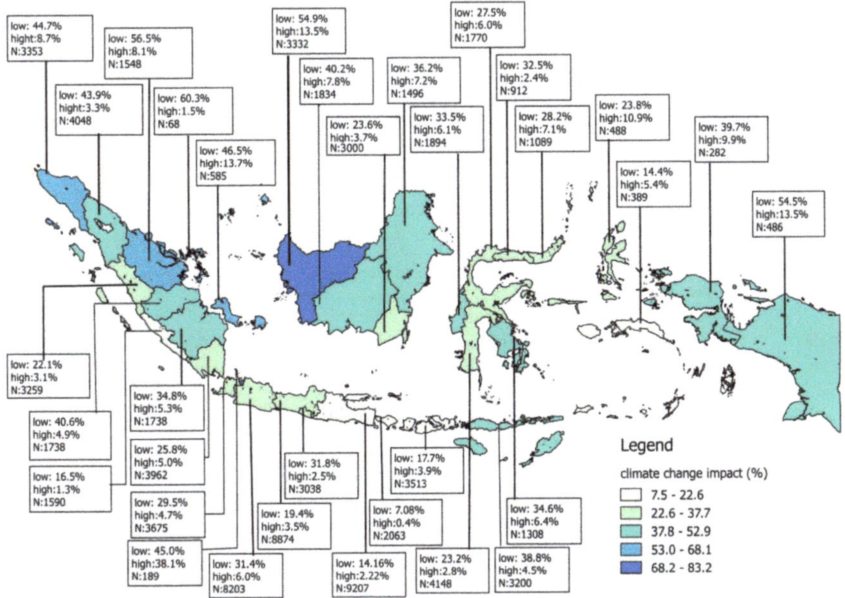

Figure 2. The distribution of the sample and the percentage of farmers who experienced effects of climate change (CC) for each province in Indonesia.

Table 1 provides summary statistics, the expected signs of predictors, and the definition of each variable. The outcome variable is the perceived impact severity of CC, which was measured by the question: "Were you impacted by climate change? If yes, how much yield loss was caused by it?" The responses consisted of 4 choices: <25% yield loss, 26–50% yield loss, 51–75% yield loss, and 75% yield loss. The response was then converted into three categories: "no impact" for those who answered 'no' to the first question, "low impact" for those who reported less than 50% yield loss, and "high impact" for those who reported greater than 50% yield loss. The purpose of this further recoding was to clearly distinguish between low and high impact. The original impact category (consisting of 4 groups) yielded a biased estimation between higher-low and lower-low impact and between higher-high and lower-high impact. The data show that 66.2% (57,771 farmers) reported no impact, 29% (25,306 farmers) reported a low impact, and 4.9% (4253 farmers) reported a high impact.

The predictors included 17 variables, which were described in Section 3. The average age of Indonesian rice farmers is 49.5 years, having only elementary schooling (5.81 years, six years of elementary school), and most are male. The majority of farmers (70.7%) cultivate their land and have an average landholding of 0.467 ha. Of rice farmers, 90% financed their farming, leaving only 9% who financed their farm with a loan. The irrigation infrastructure covered 45.3% of the rice farmers in Indonesia, whereas the rest depended on non-technical (46.4%) and rain-fed irrigation (7.9%). Ninety-six percent of farmers only cultivated rice and only 3.9% of farmers cultivated mixed species. Chemical fertilizer is the primary input in rice farming; 91% of farmers apply it, while 8.7% of farmers do not. The average cultivation frequency is twice annually. As with the institutional factors, 52% of farmers participate in a farmer group, 25% have access to extension services, and only 11.2% have participated in a farmer field school. Finally, the climatic variables showed that 7.6% of farmers experienced a flood, 18.9% of farmers experienced a drought, and 5.8% and 1.5% reported experiencing heavy rain and other climate-induced hazards, respectively.

Table 1. Summary statistics, the expected signs of predictors, and the definition of variables.

Variable	Definition	Mean and Frequency	S.D.	Expected Sign
Response Variable				
Perceived impact of CC	Ordered dummy variable (0 = farmer perceives no impact of CC; 1 = farmer perceives a low impact of CC, production decreases by ≤50%; 2 = farmer perceives a high impact of CC, production decreases by >50%)	0:57,771 (66.2%) 1:25,306 (29%) 2:4253 (4.9%)	0.579	
Social Variables				
Age	The age of the farm household head (year)	49.39 years	11.99	−
Education	The years of formal training of the farm household head (year)	5.81 years	4.20	−
Gender	Dummy variable (1 = male, 0 = female)	1:77,094 (88.3%) 0:10,236 (11.7%)	0.322	−
Economic Variables				
Land tenure	Dummy variable (1 = owned land, 0 = other)	1:61,784 (70.7%) 0:25,139 (28.8%)	0.453	−
Landholding	The area of cultivated land (ha)	0.47 ha	0.51	+
Capital source	A dummy variable, the source of used farm budget (1 = bank, 0 = other)	1:79,264 (90.8%) 0:8066 (9.2%)	0.290	−
Technical Variables				
Irrigation	A dummy variable, the type of irrigation infrastructure (1 = technical irrigation infrastructure, 0 = other)	1:39,530 (45.3%) 0:47,800 (54.7%)	0.498	−
Cropping system	A dummy variable, the type of cropping system applied (1 = monoculture farming, 0 = mixed-species farming)	1:83,942 (96.1%) 0:3388 (3.9%)	0.193	+
Fertilizer	A dummy variable, the application of chemical fertilizer by the farmer (1 = use chemical fertilizer, 0 = does not use chemical fertilizer)	1:79,744 (91.3%) 0:7586 (8.7%)	0.282	+
Cultivation frequency	The amount of rice cultivation in a year	1.64	0.696	+
Institutional Variables				
Farmer group	Dummy variable, participation in a farmer group (1 = participate, 0 = do not participate)	1:45,730 (52.4%) 0:41,600 (47.6%)	0.499	−
Extension services	A dummy variable, access to extension services (1 = having access to extension services, 0 = do not having access to extension services)	1:21,902 (25.1%) 0:65,428 (74.9%)	0.433	−
Farmer field school	A dummy variable, access to farmer field school (1 = having access to FFS, 0 = do not having access to FFS)	1:9762 (11.2%) 0:77,568 (88.2%)	0.315	−
Climatic Factors				
Flood	A dummy variable, experienced a flood (1 = experienced, 0 = did not experience)	1:6635 (7.6%) 0:80,695 (92.4%)	0.264	+
Drought	A dummy variable, experienced a drought (1 = experienced, 0 = did not experience)	1:16,538 (18.9%) 0:70,792 (81.1%)	0.392	+
Heavy rain	A dummy variable, experienced a heavy rain (1 = experienced, 0 = did not experience)	1:5069 (5.8%) 0:82,261 (94.2%)	0.233	+
Other hazards	A dummy variable, experienced other hazards (1 = experienced, 0 = did not experience)	1:1317 (1.5%) 0:8601 (98.5%)	0.121	+
Region	A regional dummy variable (1 = Sumatera, 2 = Java, 3 = Bali and Nusa Tenggara, 4 = Kalimantan, 5 = Sulawesi, 6 = Maluku and Papua)	1:23,379 (27.0%) 2:33,003 (38.2%) 3:8718 (10.1%) 4:9625 (11.1%) 5:10,556 (12.2%) 6:1156 (1.3%)		

Note: − and + denote decreasing and increasing effect to farmers' P-I, respectively.

3.3. Empirical Model

To estimate the effect of each predictor on the ordinal response variable, we used an ordered probit regression. Ordered probit estimate can be used to estimate how each predictor determines the probability that farmers perceive (whether high or low) an impact of CC on their rice farming. The use of ordered-probit regression to analyze the effect of independent variables on the ordinal response was favored to avoid false alarm (detecting a non-existent effect) and loss of power (failure to detect an effect) problems [83]. Equation (1) specifies the model:

$$y_i^* = \sum_{i=1}^{17} \beta x_i + \varepsilon_i, i = 1, 2, \dots, N \tag{1}$$

where y_i^* is the response variable that represents the perceived impact of CC, β is the parameter to be estimated, x_i is the vector of predictors, ε_i is the error term, and N is the number of observations.

4. Result and Discussion

4.1. Estimation Results

Of the 17 predictors analyzed in the ordered probit regression, 13 had a statistically significant effect on the perceived impact of CC, 10 of these had the expected signs, 6 variables had a positive sign, and 7 had a negative sign. However, since the climatic variables logically increase the degree of the P-I, nine variables practically represent farmer characteristics that have a statistically significant effect. The likelihood test ratio and the pseudo R^2 indicated that the model is robust.

The estimation results of the social variables revealed that education and gender have a statistically significant effect on P-I, having negative and positive effects, respectively. In the economic category, only land tenure had a statistically significant adverse effect. The technical variables seem to primarily affect the degree of P-I since all variables in this category have a statistically significant effect. Farmers with access to technical irrigation, practicing monoculture farming, and applying chemical fertilizer reported a lower degree of P-I. However, farmers with a higher cultivation frequency seem to have experienced a greater impact of CC. The results of institutional variables suggest that participation in a farmer group and having access to extension services reduce the degree of P-I. However, participation in a farmer's field school is not likely to have a significant effect in terms of decreasing the degree of P-I.

The purpose of incorporating climatic variables in the estimation was to identify which type of climate-related hazard is perceived as causing the most severe damage. The estimation results revealed that all variables in this category have a statistically significant positive effect on the degree of P-I. The obtained coefficients indicated that farmers perceive flood as causing the most damage, followed by drought, heavy rain, and other hazards. The estimation result for each variable is provided in Table 2.

Table 2. The estimation results.

Variable Name	Estimate	Sig.
Response Variable		
Perceived impact of CC		
Threshold *low impact* (1)	−24.133	0.000 ***
Threshold *high impact* (2)	−19.532	0.000 ***
Social Variables		
Age	−0.001	0.513 [ns]
Education	−0.016	0.000 ***
Gender (Male)	0.071	0.011 **
Economic Variables		
Land tenure (Own land)	−0.039	0.050 **
Landholding	−0.002	0.907 [ns]
Capital source (loan)	0.030	0.340 [ns]
Technical Variables		
Irrigation (technical irrigation)	−0.075	0.000 ***
Cropping system (monoculture)	−0.324	0.000 ***
Fertilizer (applying fertilizer)	−0.098	0.000 ***
Cultivation frequency	0.030	0.042 **
Institutional Variables		
Farmer group	−0.034	0.091 *
Extension services	−0.060	0.019 **
Farmer field school	−0.001	0.980 [ns]
N	87,330	

Table 2. *Cont.*

Variable Name	Estimate	Sig.
Climatic Factors		
Flood	7.271	0.000 ***
Drought	7.086	0.000 ***
Heavy rain	6.746	0.000 ***
Other hazards	6.682	0.000 ***
Regional Variables		
Sumatera	−0.577	0.000 ***
Java	−0.394	0.000 ***
Bali and Nusa Tenggara	−0.549	0.000 ***
Kalimantan	−0.443	0.000 ***
Sulawesi	−0.472	0.000 ***
Maluku and Papua	−0.512	0.000 ***
Regression Robustness		
Likelihoodtest ratio	135,205.866	0.000 ***
Pearson goodness of fit	28,051.967	1.000 [ns]
Deviance goodness of fit	22,696.188	1.000 [ns]
Cox and Snell R^2	0.789	
Nagelkerke R^2	0.998	
N	87,330	

Note: ***, **, and * denote significant at 1%, 5%, and 10% levels, respectively. [ns] denotes a statistically not significant effect.

4.2. Discussion

The identification of factors affecting the degree of a farmer's P-I of CC was the primary purpose of this study. The estimation results in Table 2 show the effect of each variable. The variables were further classified into weakening and reinforcing factors. A reinforcing factor decreases the degree of P-I, whereas a weakening factor increases it. Figure 3 summarizes the reinforcing and weakening factors.

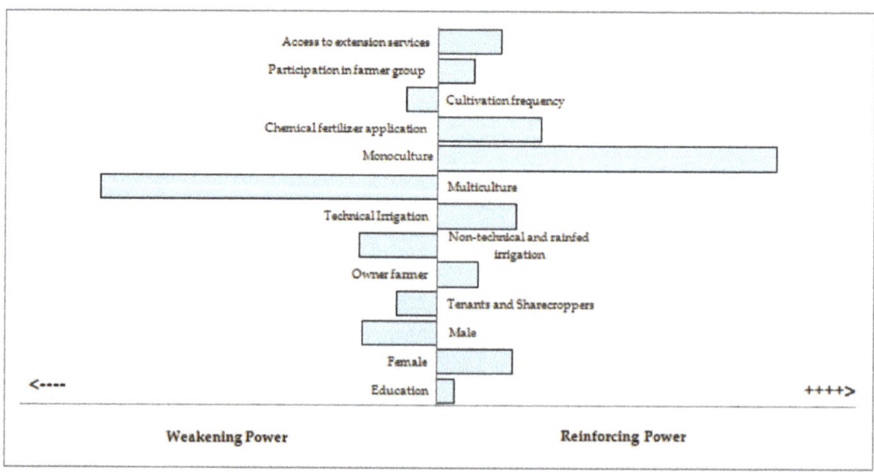

Figure 3. The reinforcing and weakening factors of the climate change perceived impact.

4.2.1. Social Variables

The estimation results show that two out of three variables had a statistically significant effect. Education and gender had significantly negative and positive effects, respectively, whereas age did not have any statistically significant effect. The result suggests that farmers with higher levels of education better adapt to CC and thus perceive a lower degree of CC impact. Many studies have reported the role of education in moderating the adverse effects of CC. A study on farmer use of CC adaptation practices in Pakistan revealed that farmers with higher education are better at adjusting sowing time, using a drought-tolerant crop, and practicing crop-shifting, and these practices have resulted in higher food security [39]. Another study in Pakistan [67] and one in Kenya [84] that assessed farmer awareness of CC also found a positive effect of education on adaptation ability. Similarly, a study in Ethiopia found that education positively affects the probability of farmers adapting to CC via soil conservation and changing planting dates [65]. These findings show that farmers with higher education perceive a lower degree of CC impact because they are better at adapting to and have a higher awareness of CC.

The positive regression coefficient of gender indicates that male farmers perceived a higher degree of impact than female farmers. The gender issue in the CC discussion has gained considerable attention. In some studies, male farmers were reported to be better at adapting to CC. A higher likeliness to undertake adaptation practices was found among male farmers in Ethiopia [65], and a higher awareness of CC was found among male farmers in Kenya [84]. However, in a Pakistan study, female farmers were found to be better at using adaptation practices and had a higher level of food security [39]. A cross-European analysis of individual perceptions of CC revealed that women in most European countries are more aware of CC, but the degree of awareness varied across countries [44]. Also, female representation in the national parliaments is related to the creation of stringent climate change policies and lower CO_2 emissions [85]. Furthermore, the theory of socialization stated that female possesses stronger cooperation and carefulness, personal traits that are relevant to the success of CC action, than male [86].

This finding suggests that the gender effect is context- and location-specific. In this study, female farmers perceived a lower degree of impact. Based on the data, female farmers participate more in crop insurance than male farmers. The participation rate of female farmers in crop insurance is 0.255%, whereas that of the male farmer is 0.192%. Women farmers have a more positive perception regarding future farming conditions. Figure 4 shows the distribution of farmer perception of future farming conditions.

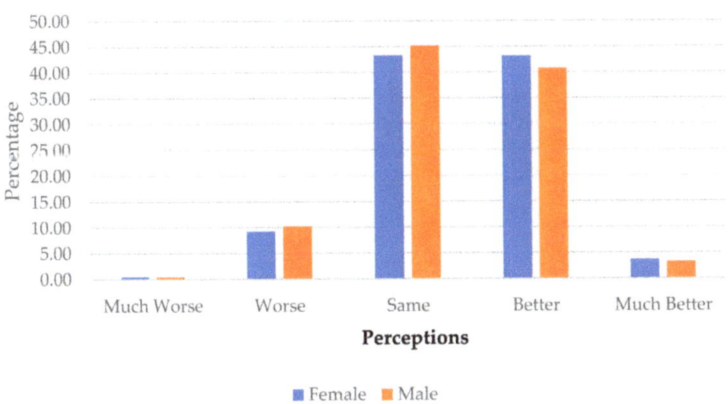

Figure 4. Perception of female and male farmers on future farming conditions.

The regression result revealed that age has no statistically significant effect on the degree of P-I. Age has often been associated with farming experience and increases the adoption of adaptation practices [65] and awareness of CC [67,84]. However, another study suggested that younger farmers are more likely to adopt adaptation practices [39]. A cross-European study obtained mixed results in terms of the age effects on individual perception of CC [44]. Thus, the results in this study and of the previous study indicate that, similar to gender, the age effect is location- and context-specific.

4.2.2. Economics Factors

The results of this study revealed that land tenure decreases the degree of perceived climate change impact. Landholding and farm capital source did not have statistically significant effects. The effect of land tenure security on the rate of adaptation practices varies. A study in Ghana reported that land ownership increases a farmer's likeliness to undertake adaptation practices [87]. Conversely, another study in Pakistan found that land ownership reduces it [88]. This study suggests that other characteristics, such as education, household size, and resource access, might influence the identified adverse effect of land ownership. Similarly, a study in Pakistan found an adverse effect of land ownership, even after controlling for land size and wealth-related characteristics, but not for education, which was found to have a strong positive effect [39]. These findings suggest that a stronger variable, such as education, might offset the effect of land ownership. The results of this study indicate that land ownership decreases the degree of P-I of CC.

Since we analyzed the P-I of CC on crop yield, the result indicates that farm owners are more willing to adopt yield-enhancing adaptation strategies than farm tenants and sharecroppers. Farm owners have a higher incentive to invest in adaptation practices [89]. A study in Pakistan and Indonesia found that insecure land tenure decreases adaptation practices [90] and makes farmers less cautious in terms of their farming, manifesting, for example, through an absence of soil conservation [91]. The positive effect of secure land tenure on adaptation practices, however, has potential disadvantages when the adaptation yielded unintended maladaptive outcomes. For example, a study on tomato growers in Ghana revealed that the perception of climate variability drives farmer to use more agrochemicals to retain crop production, and these chemicals cause the pollution of nearby water bodies and increase soil acidity above the optimum crop requirements [18]. Similar practices have been identified in rural areas of Indonesia, where watermelon farmers use excessive chemical pesticides to retain production, causing groundwater pollution that damages the quality of water for consumption [92].[1] Ultimately, adaptation becomes maladaptation, which erodes sustainable development and shifts vulnerability to other actors. Thus, it is essential to guide farmer's adaptation practices to avoid unintended maladaptive outcomes.

4.2.3. Technical Factors

The probit estimation revealed that each variable in the technical category has a statistically significant effect. The regression coefficient of the irrigation variable indicated that farmers with access to technical irrigation perceive a lower degree of impact than farmers with non-technical and rain-fed irrigation. Irrigation infrastructure is crucial for reducing the adverse effects of CC. The availability of irrigation infrastructure (II) increases the efficiency of the distribution of limited water and decreases crop yield loss due to drought. In some cases, II becomes a drainage facility that reduces crop damage caused by flood. The establishment and improvement of II is a key framing device for CC adaptation in a rice-based country such as Vietnam [94]. In Vietnam, II reduces water availability during the rainy season and increases it during the dry season [95].

Technical irrigation in Indonesia has been developed since the 1970s with the support of the Green Revolution program. The establishment of technical irrigation infrastructure has increased the yields, cropping season, and cropping intensity of rice farming in Indonesia [73]. Conversely, both non-technical and rain fed agriculture rely mainly on rainfall to supply irrigation. Hence, those in such areas are vulnerable to drought occurrence and pay a higher cost of irrigation due to water scarcity. Rice and agricultural productivity in this area are lower compared to areas with technical irrigation, which means that farmers such areas are economically more vulnerable to CC.

The fertilizer variable demonstrated that the application of chemical fertilizer decreases the degree of P-I. In line with the previous discussion, farmers often apply more agrochemicals to retain their crop

---

[1]   The example of data on extensive use of agrochemical in Indonesian farming can be found here [93].

production. A study in Nepal identified that farmers adapt to CC by applying more chemical fertilizer and this increases rice productivity [9]. This finding suggests that autonomous adaptation will likely become maladaptive if a broader perspective is included. This is not surprising as the primary goal of farmers is to increase crop yield or limit crop yield loss, and, in most cases, this is also the priority of agricultural policy. Thus, a further adaptation policy should guide farmers to implement appropriate adaptation practices to reduce unintended maladaptive outcomes.

The results also indicate that monoculture farming decreases the degree of perceived CC impact. Monoculture is often regarded as a threat to biodiversity. Thus, multiculture farming is commonly employed to reduce biodiversity loss in trees farming. However, monoculture in rice farming is less of a threat to biodiversity since its temporal dimension is short [96], and the majority of farmers practice crop rotation. The last variable, cultivation frequency, increases the degree of P-I. A higher cultivation frequency prolongs exposure to CC, which increases the probability of farmers being affected by CC and increases the degree of P-I.

4.2.4. Institutional Factors

The estimation results show that access to extension services and participation in a farmer group significantly decrease the degree of P-I, whereas participation in a farmer's field school does not. Extension services are effective information channels used to raise farmer awareness of CC and drive adaptation practices. The importance of extension services as a means of delivering accurate information about CC has been addressed in many studies. A study of farm households in four provinces in Pakistan demonstrated that access to extension services increases how likely a farmer is to adapt to CC [39]. Similarly, a study on rice farmers in the Terai and Hill area of Pakistan indicated that farmers who receive information from extension agents are more likely to adapt [9]. Extension services have been the focus of studies in East Africa [97], Ethiopia [65], and Punjab, Pakistan [67], to enhance a farmer's resilience against CC. These finding indicate that information is crucial in shaping and directing a farmer's behavior with respect to CC. Access to timely and accurate information not only increases a farmer's awareness and adaptation to CC, but might also be used to inform farmers about adaptation practices that yield maladaptive outcomes.

Indonesian farmers receive extension services from several sources, such as state agricultural extension officers (PPL), a state pest-control officer (POPT), an officer from the office of agriculture at the district level (Diperta), and private extension services. Figure 5 shows the number of farmers who receive extension services based on the type of extension service officer. Only 25% of rice farmers in Indonesia have access to these services. The primary extension agent is from the government. The data shows that non-government extension agents vary in type, and these may be from a private corporation (such as a pesticide and fertilizer factory), a non-governmental organization (NGO), or may be associated with peer-farmer extension. Increasing the number and coverage of extension agents is essential for CC adaptation policy in Indonesia, since the majority of farmers (41%) receiving extension services are concentrated in Java, with 21.4% in Sumatera, 8.3% in Bali and Nusa Tenggara, 10.6% in Kalimantan, 14.4% in Sulawesi, and 1.2% and 1.9% in Maluku and Papua, respectively.

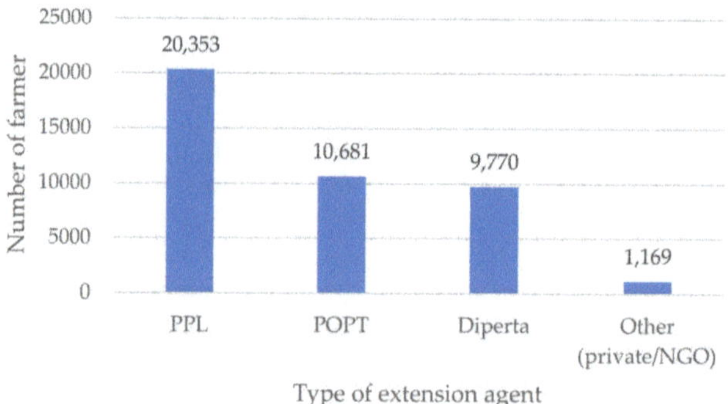

Figure 5. The number of farmers who receive extension services based on the type of extension agents.

The government should encourage the establishment of farmer-to-farmer extension services given the potential human resource in Indonesian rice farming. Currently, 82% of the rice farmer population is 20–60 years of age, and 2% of rice farmers have a bachelor degree or higher. Of this educated group (bachelor degree or higher), 87% are aged between 20 and 60 years, with 25% in 20–40 years, and 61% in the 41–60 years old group. Facilitating the establishment of farmer-to-farmer extension provides a strong foundation for sustainable agricultural extension. Farmer-to-farmer extension provides a cost-effective training method for farmers, since trained farmers disseminate their useful knowledge to the non-trained farmer, and the non-trained farmer is eager to adopt the technology used by the trained farmer to increase yield and profit [98]. A strong positive effect of formal education on a farmer's resilience with respect to CC suggests that farmer-to-farmer extension provided by an educated farmer potentially increases the resilience of the recipient.

The estimation results indicate that participation in a farmer group decreases the degree of P-I. Participation in a farmer group is the primary requisite for obtaining farm inputs, extensions, and other programs provided by the government [79]. Several studies have reported the importance of farmer groups in increasing farmer adaptation to CC [65,97], awareness to CC [67], and farm productivity and food security [39].

4.2.5. Climatic and Regional Factors

This category focuses on identifying the type of impact that a farmer perceives as most severely damaging their farming. The category contains four variables: Flood, drought, heavy rain, and other hazards (such as landslides). The estimation results show that farmers perceive floods as causing the most severe impact, followed by drought, heavy rain, and other hazards. Flood and drought are the primary consequences of changes in rainfall frequency and intensity. The occurrences of drought and flood are the cause of agricultural yield loss in developing countries and generally affect large areas [99,100]. Figure 6 shows the distribution of farmers who have experienced flood and drought events in Indonesia. The distribution shows that more farmers have experienced droughts than they have floods. Even in provinces with high rainfall intensity, droughts are more frequent than floods. The spatial distribution of affected farmers shows that farmers outside Java are more vulnerable to CC. This information is vital in the implementation of adaptation policies. The government should prioritize the increase in the climate resiliency of farmers outside of Java Island. The increase in resiliency can be achieved by expanding the coverage of agricultural extension services to farmers.

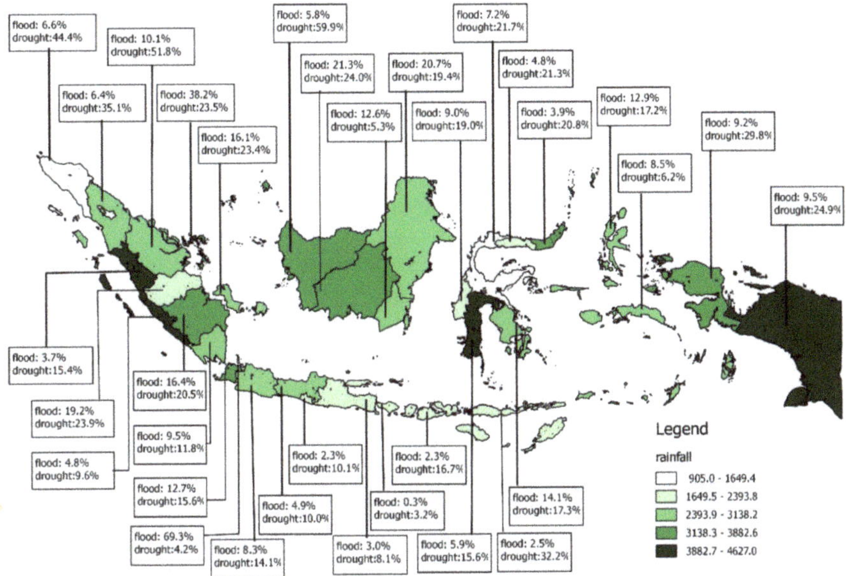

Figure 6. The distribution of farmers experiencing flood and drought in each province of Indonesia (the rainfall intensity unit is mm/year).

The regional variables demonstrate regions whose farmers perceive a higher degree of CC. The estimation results indicate that farmers in Sumatera are likely to perceive the impact of CC. Twenty-seven percent of Indonesian rice farmers are in Sumatera, so a large impact of CC on rice farmers in Sumatera is perceived, since the national rice supply significantly decreased. Thirty-eight percent of Indonesian rice farmers in Java perceive a low impact of CC. The agricultural infrastructure in Java is more advanced than other regions in Indonesia. Physical and institutional infrastructure have been established in Java since the 1970s. Thus, in the future, it is critical to emphasize agricultural regions outside Java. The order of regions from the most vulnerable to the most resilient is as follows: Sumatera, Bali and Nusa Tenggara, Maluku and Papua, Sulawesi, and Kalimantan.

4.2.6. Policy Implications

The primary purpose of an agricultural CC adaptation policy is to reverse its adverse effects. Adaptation to CC limits agricultural yield loss and improves food security. In some cases, CC is beneficial to agricultural production if proper adaptation practices are implemented [101,102]. Whereas the majority of countries have formulated a specific adaptation strategy to be implemented at the farm level, the remaining challenge is delivering the strategy so that it is adopted by many farmers. The theory of risk perception suggests that individuals who perceive a higher degree of risk would be willing to take any action required to remove the risk [27–29]. In the case of the CC impact on agriculture, farmers will be more willing to adopt adaptation strategy if they individually perceive CC damage. However, this high likeliness to adapt becomes a problem if a farmer's autonomous adaptation creates maladaptive outcomes. Thus, identifying the determinants of a farmer's P-I is crucial in the implementation of adaptation policy. This information can be used to increase the rate of adaptation practices and limit maladaptation.

The results of this study show that the P-I of CC is affected by various factors. Female farmers are shown to be more resilience in managing the impact of CC than male farmers. The theory of socialization suggests that female possesses a stronger personal traits that are relevant to CC action (mitigation and adaptation) than male. Hence, increasing female farmer participation in adaptation activity and in the CC-related decision-making process is crucial to increase farmer resilience toward CC. Formal education and access to extension services have a strong effect in terms of reducing the degree of P-I, and this strongly agrees with previous studies in many developing countries. Therefore, information is crucial to increasing a farmer's resilience with respect to CC. Thus, we suggest strengthening information about CC as a primary policy. Information about CC can be strengthened in two aspects: First, by increasing the coverage of current extension services and, second, by establishing farmer-to-farmer extension, where educated farmers (those with a bachelor degree or higher) provide education to their peers. Among other factors, land tenure is crucial in adaptation policy. Findings in this study suggest that the security of tenure drives farmers to provide the investment and pay for the running costs of adaptation. Economically, the security of tenure ensures that farmers are incentivized to adopt of adaptation practices.

1. Policy implications that are suggested based on these findings are as follows: Female farmer are better at managing the CC impact. It is essential to increase female farmer participation in CC action (adaptation and mitigation) and in the CC-related decision making process. Additionally, extension services are currently an effective method of delivering the substance of a policy to the farmer. It is crucial to use current channels of agricultural extension to provide adaptation strategies to farmers and to expand the coverage area of the extension services.

2. The government should facilitate farmer-to-farmer extension, which can be implemented by identifying key farmers (farmers with a high degree of formal education), providing them with intensive training, and facilitating the dissemination of their expertise. This strategy is feasible because Indonesian rice farmers with a high degree of formal education are aged between 20 and 40 years and have a high potential to be a key farmer in this framework.

3. Efforts should be made to implement adaptation policies based on a farmer's land tenure and prioritize farmers who cultivate land under a lease or sharecropping contract. The security of tenure affects a farmer's incentive to adopt and conduct adaptation practices. The farmer who cultivates land under a lease or sharecropping contract will put little effort and investment into adaptation practices. Currently, 20% of Indonesian rice farmers are in this category; if they do not practice adaptation strategies, the decrease in rice production will be substantial. In the context of food security, the national rice supply will decrease.

5. Conclusions

We attempted to identify factors affecting the degree of P-I of CC on farm yield among rice farmers in Indonesia. The P-I is a subjective measure of the impact of CC on farm yield. This subjective nature of P-I is essential because it indicates how farmers will adapt to CC. The more severe the P-I of CC, the more likely farmers are to adopt adaptation practices. In general, the results seem to support the findings of previous empirical studies and that little difference exists between the actual and perceived impact of CC. Higher education, secure land tenure, the existence of irrigation infrastructure, and access to extension services decrease the degree of the P-I of CC. Previous empirical studies showed that these variables improve how likely a farmer is to adopt CC adaptation practices. Since farmers with a high value for these variables are likely to undertake adaptation practices, we conclude that adaptation limits the adverse effects of CC on farm yield. Hence, the farmer perceives a lower impact of CC.

However, since we measured the perceived impact of CC on farm yield, the existence of maladaptation is suggested. A farmer's primary objective is to retain their yield level, and it is likely that adaptations are primarily aimed to limit yield loss. Consequently, such adaptation potentially creates maladaptive outcomes. In Section 4.2.3, the application of chemical fertilizer was found to decrease the degree of P-I. This suggests that farmers might use excessive chemical fertilizer (inputs) to reduce yield loss. This excessive use will pollute nearby water bodies and decrease soil quality.

Finally, the information obtained from this study is nationally representative and is relevant for the National Adaptation Policy in Indonesia. However, the current study is limited in providing the details of adaptation practices (the type and form of adaptation) in which farmers engage. Thus, additional study that captures the detailed conditions and practices of Indonesian rice farming is crucial in supporting the information in this study.

Thus, our results provide a basis for future research directions:

1. As extension services play a vital role in providing information to other farmers, an investigation into the mechanism by which extension services improve the resilience of farmers with respect to CC will provide essential information for improving the efficiency of extension services in delivering timely information to the farmer.
2. The identification of which type of extension officer and what information contributes most to increasing farmer resilience with respect to CC will yield vital information for increasing overall resilience to CC.
3. The identification of the type and form of adaptation practices among Indonesian rice farmers and their outcome is crucial for identifying whether a farmer's practices will lead to appropriate adaptation or maladaptive outcomes.

Author Contributions: Conceptualization: M.R., A.K., Y.M. and T.K.; formal analysis: M.R., A.K. and Y.M.; funding acquisition: T.K.; investigation: Y.M.; resources: Y.M. and T.K.; software: M.R. and A.K.; supervision: M.R. and T.K.; visualization: M.R. and A.K.; writing—original draft: M.R. and A.K.; writing—review & editing: M.R.

Funding: This research was funded by the Japan Society for the Promotion of Science, Grant Number: 18K05839.

Conflicts of Interest: The authors declare no conflict of interest.

References

1. Rojas-Downing, M.M.; Nejadhashemi, A.P.; Harrigan, T.; Woznicki, S.A. Climate change and livestock: Impacts, adaptation, and mitigation. *Clim. Risk Manag.* **2017**, *16*, 145–163. [CrossRef]
2. Rust, J.M. The impact of climate change on extensive and intensive livestock production systems. *Anim. Front.* **2019**, *9*, 20–25. [CrossRef]
3. McCarl, B.A.; Hertel, T.W. Climate change as an agricultural economics research topic. *Appl. Econ. Perspect. Policy* **2018**, *40*, 60–78. [CrossRef]
4. Knox, J.; Hess, T.; Daccache, A.; Wheeler, T. Climate change impacts on crop productivity in Africa and South Asia. *Environ. Res. Lett.* **2012**, *7*. [CrossRef]
5. IPCC. *Climate Change 2014: Impacts, Adaptation, and Vulnerability. Part A: Global and Sectoral Aspects.* 2014. Available online: http://wedocs.unep.org/handle/20.500.11822/19908 (accessed on 30 March 2019).
6. Jamshidi, O.; Asadi, A.; Kalantari, K.; Azadi, H.; Scheffran, J. Vulnerability to climate change of smallholder farmers in the Hamadan province, Iran. *Clim. Risk Manag.* **2019**, *23*, 146–159. [CrossRef]
7. Seo, S.N. An analysis of public adaptation to climate change using agricultural water schemes in South America. *Ecol. Econ.* **2011**, *70*, 825–834. [CrossRef]
8. Stage, J. Economic valuation of climate change adaptation. *Ann. N. Y. Acad. Sci.* **2010**, *1185*, 150–163. [CrossRef]
9. Khanal, U.; Wilson, C.; Hoang, V.N.; Lee, B. Farmers' Adaptation to Climate Change, Its Determinants and Impacts on Rice Yield in Nepal. *Ecol. Econ.* **2018**, *144*, 139–147. [CrossRef]
10. Waha, K.; Müller, C.; Bondeau, A.; Dietrich, J.P.; Kurukulasuriya, P.; Heinke, J.; Lotze-Campen, H. Adaptation to climate change through the choice of cropping system and sowing date in sub-Saharan Africa. *Glob. Environ. Chang.* **2013**, *23*, 130–143. [CrossRef]
11. Adger, W.N.; Agrawala, S.; Mirza, M.M.Q.; Conde, C.; O'Brien, K.; Pulhin, J.; Pulwarty, R.; Smit, B.; Takahashi, K. Assessment of adaptation practices, options, constraints and capacity. In *Climate Change 2007*; Parry, M.L., Canziani, O.F., Palutikof, J.P., Hanson, C.E., van der Linden, P.J., Eds.; Cambridge University Press: Cambridge, UK, 2007; pp. 719–743.
12. Hossain, K.; Quaik, S.; Ismail, N.; Rafatullah, M.; Ali, I. Climate Change-Perceived Impacts on Agriculture, Vulnerability and Response Strategies for Improving Adaptation Practice in Developing Countries (South Asian Region). *Int. J. Agric. Res.* **2016**, *11*, 1–12. [CrossRef]

13. Makuvaro, V.; Walker, S.; Masere, T.P.; Dimes, J. Smallholder farmer perceived effects of climate change on agricultural productivity and adaptation strategies. *J. Arid Environ.* **2018**, *152*, 75–82. [CrossRef]
14. Gawith, D.; Hodge, I. Moving beyond description to explore the empirics of adaptation constraints. *Ecol. Indic.* **2018**, *95*, 907–916. [CrossRef]
15. Hasan, M.K.; Kumar, L. Comparison between meteorological data and farmer perceptions of climate change and vulnerability in relation to adaptation. *J. Environ. Manage.* **2019**, *237*, 54–62. [CrossRef] [PubMed]
16. Neset, T.S.; Wiréhn, L.; Klein, N.; Käyhkö, J.; Juhola, S. Maladaptation in Nordic agriculture. *Clim. Risk Manag.* **2019**, *23*, 78–87. [CrossRef]
17. Juhola, S.; Glaas, E.; Linnér, B.O.; Neset, T.S. Redefining maladaptation. *Environ. Sci. Policy* **2016**, *55*, 135–140. [CrossRef]
18. Guodaar, L.; Asante, F.; Eshun, G.; Abass, K.; Afriyie, K.; Appiah, D.O.; Gyasi, R.; Atampugre, G.; Addai, P.; Kpenekuu, F. How do climate change adaptation strategies result in unintended maladaptive outcomes? Perspectives of tomato farmers. *Int. J. Veg. Sci.* **2019**, *00*, 1–17. [CrossRef]
19. Work, C.; Rong, V.; Song, D.; Scheidel, A. Maladaptation and development as usual? Investigating climate change mitigation and adaptation projects in Cambodia. *Clim. Policy* **2018**, *0*, 1–16. [CrossRef]
20. Parry, J.-E.; Terton, A. How Are Vulnerable Countries Adapting to Climate Change—Frequently Asked Questions (FAQ's). Available online: https://www.iisd.org/faq/adapting-to-climate-change/ (accessed on 20 Feburary 2019).
21. Bandara, J.S.; Cai, Y. The impact of climate change on food crop productivity, food prices and food security in South Asia. *Econ. Anal. Policy* **2014**, *44*, 451–465. [CrossRef]
22. Mendelsohn, R. The Impact of Climate Change on Agriculture in Developing Countries The Impact of Climate Change on Agriculture in Developing Countries. *J. Nat. Resour. Policy Res.* **2008**, *1*, 5–19. [CrossRef]
23. Barbier, E.B.; Hochard, J.P. The impacts of climate change on the poor in disadvantaged regions. *Rev. Environ. Econ. Policy* **2018**, *12*, 26–47. [CrossRef]
24. Mase, A.S.; Gramig, B.M.; Prokopy, L.S. Climate change beliefs, risk perceptions, and adaptation behavior among Midwestern U.S. crop farmers. *Clim. Risk Manag.* **2017**, *15*, 8–17. [CrossRef]
25. Weber, E.U. What shapes perceptions of climate change? *Wiley Interdiscip. Rev. Clim. Chang.* **2010**, *1*, 332–342. [CrossRef]
26. Weber, E.U. Experience-based and description-based perceptions of long-term risk: Why global warming does not scare us (yet). *Clim. Chang.* **2006**, *77*, 103–120. [CrossRef]
27. Renn, O.; Burns, W.J.; Kasperson, J.X.; Kasperson, R.E.; Slovic, P. The Social Amplification of Risk: Theoretical Foundations and Empirical Applications. *J. Soc. Issues* **1992**, *48*, 137–160. [CrossRef]
28. Slovic, P.; Fischhoff, B.; Lichtenstein, S. Why Study Risk Perception? *Risk Anal.* **1982**, *2*, 83–93. [CrossRef]
29. Slovic, P. Informing and Educating the Public About Risk. *Risk Anal.* **1986**, *6*, 403–415. [CrossRef] [PubMed]
30. Nordhaus, W.D. To Slow or Not to Slow: The Economics of The Greenhouse Effect. *Econ. J.* **1991**, *101*, 920. [CrossRef]
31. Cline, W.R. *Global Warming and Agriculture: Impact Estimates by Country;* Center for Global Development & Peterson Institute for International Economics: Washington, DC, USA, 2007; ISBN 978-0-88132-403-7.
32. Mendelsohn, R.; Dinar, A.; Sanghi, A. The effect of development on the climate sensitivity of agriculture. *Environ. Dev. Econ.* **2001**, *6*, 85–101. [CrossRef]
33. Mendelssohn, R.; Dinar, A. Climate change, agriculture, and developing countries: Does adaptation matter? *World Bank Res. Obs.* **1999**, *14*, 277–293. [CrossRef]
34. Kumar, K.S.K.; Parikh, J. Indian agriculture and climate sensitivity. *Glob. Environ. Chang.* **2001**, *11*, 147–154. [CrossRef]
35. Sarker, M.A.R.; Alam, K.; Gow, J. Assessing the effects of climate change on rice yields: An econometric investigation using Bangladeshi panel data. *Econ. Anal. Policy* **2014**, *44*, 405–416. [CrossRef]
36. Xie, W.; Huang, J.; Wang, J.; Cui, Q.; Robertson, R.; Chen, K. Climate change impacts on China's agriculture: The responses from market and trade. *Chin. Econ. Rev.* **2018**. [CrossRef]
37. Chalise, S.; Naranpanawa, A.; Bandara, J.S.; Sarker, T. A general equilibrium assessment of climate change–induced loss of agricultural productivity in Nepal. *Econ. Model.* **2017**, *62*, 43–50. [CrossRef]
38. Al-Amin, A.Q.; Ahmed, F. Food Security Challenge of Climate Change: An Analysis for Policy Selection. *Futures* **2016**, *83*, 50–63. [CrossRef]

39. Ali, A.; Erenstein, O. Assessing farmer use of climate change adaptation practices and impacts on food security and poverty in Pakistan. *Clim. Risk Manag.* **2017**, *16*, 183–194. [CrossRef]

40. Elum, Z.A.; Modise, D.M.; Marr, A. Farmer's perception of climate change and responsive strategies in three selected provinces of South Africa. *Clim. Risk Manag.* **2017**, *16*, 246–257. [CrossRef]

41. Goeldner-Gianella, L.; Grancher, D.; Magnan, A.K.; de Belizal, E.; Duvat, V.K.E. The perception of climate-related coastal risks and environmental changes on the Rangiroa and Tikehau atolls, French Polynesia: The role of sensitive and intellectual drivers. *Ocean Coast. Manag.* **2019**, *172*, 14–29. [CrossRef]

42. Soto-Montes-de-Oca, G.; Alfie-Cohen, M. Impact of climate change in Mexican peri-urban areas with risk of drought. *J. Arid Environ.* **2019**, *162*, 74–88. [CrossRef]

43. Steeves, L.; Filgueira, R. Stakeholder perceptions of climate change in the context of bivalve aquaculture. *Mar. Policy* **2019**, *103*, 121–129. [CrossRef]

44. Poortinga, W.; Whitmarsh, L.; Steg, L.; Böhm, G.; Fisher, S. Climate change perceptions and their individual-level determinants: A cross-European analysis. *Glob. Environ. Chang.* **2019**, *55*, 25–35. [CrossRef]

45. Ministry of Agriculture. *Agricultural Statistics*; Susanti, A.A., Waryanto, B., Eds.; Center for Agricultural Data and Information System (Ministry of Agriculture): Jakarta, Indonesia, 2018; ISBN 9798958659.

46. BPS-Statistics Indonesia. *Statistical Yearbook of Indonesia*; Sub-Directorate of Statistical Compilation and Publication, Ed.; BPS—Statistics Indonesia: Jakarta, Indonesia, 2018; ISBN 0126-2912.

47. BPS-Statistics Indonesia. The number of farm household, cultivation area, and average cultivation area for food crops (in Indonesian: Jumlah Rumah Tangga, Luas Tanam, dan Rata-rata Luas Tanam Usaha Tanaman Padi dan Palawija menurut Jenis Tanaman). Available online: https://st2013.bps.go.id/dev2/index.php/site/tabel?tid=66&wid=0 (accessed on 4 March 2019).

48. Surmaini, E.; Runtunuwu, E.; Las, I. Efforts of Agricultural Sector in Dealing with Climate Change. *J. Litbang Pertan.* **2011**, *30*, 1–7. [CrossRef]

49. Brown, K. *Resilience, Development and Global Change*; Routledge: New York, NY, USA, 2016; ISBN 9780415663465.

50. Forsyth, T. Is resilience to climate change socially inclusive? Investigating theories of change processes in Myanmar. *World Dev.* **2018**, *111*, 13–26. [CrossRef]

51. Candradijaya, A. Pemanfaatan Model Proyeksi Iklim dan Simulasi Tanaman Dalam Penguatan Adaptasi Sistem Pertanian Padi Terhadap Penurunan Produksi Akibat Perubahan Iklim di Kabupaten Sumendang. *Inform. Pertaanian* **2014**, *23*, 159–168. [CrossRef]

52. Hidayati, I.N.; Suryanto, S. Pengaruh Perubahan Iklim Terhadap Produksi Pertanian Dan Strategi Adaptasi Pada Lahan Rawan Kekeringan. *J. Ekon. Stud. Pembang.* **2015**, *16*, 42–52. [CrossRef]

53. Putri FA, S. Strategi Adaptasi Dampak Perubahan Iklim Terhadap Sektor Pertanian Tembakau. *J. Ekon. dan Stud. Pembang.* **2012**, *13*, 30–42.

54. Van Der Kaars, S.; Dam, R. Vegetation and climate change in West-Java, Indonesia during the last 135,000 years. *Quat. Int.* **1997**, *37*, 67–71. [CrossRef]

55. Yuliawan, T.; Handoko, I. The effect of temperature rise to rice crop yield in Indonesia uses Shierary Rice model with geographical information system (GIS) feature. *Procedia Environ. Sci.* **2016**, *33*, 214–220. [CrossRef]

56. Susilowardhani, A. The Potential of Strategic Environmental Assessment to Address The Challenges of Climate Change to Reduce The Risks of Disasters: A Case Study From Semarang, Indonesia. *Procedia Soc. Behav. Sci.* **2014**, *135*, 3–9. [CrossRef]

57. Santoso, D.; Suroso, A.; Kombaitan, B.; Setiawan, B. Exploring the use of risk assessment approach for climate change adaptation in Indonesia: Case study of flood risk and adaptation assessment in the South Sumatra province. *Procedia Environ. Sci.* **2013**, *17*, 372–381.

58. Zikra, M. Climate Change Impacts on Indonesian Coastal Areas. *Procedia Earth Planet. Sci.* **2015**, *14*, 57–63. [CrossRef]

59. Petrich, C.H. Indonesia and global climate change negotiations: Potential opportunities and constraints for participation, leadership, and commitment. *Glob. Environ. Chang.* **1993**, *3*, 53–77. [CrossRef]

60. Di Gregorio, M.; Fatorelli, L.; Paavola, J.; Locatelli, B.; Pramova, E.; Ridho, D.; May, P.H.; Brockhaus, M.; Maya, I.; Dyah, S. Multi-level governance and power in climate change policy networks. *Glob. Environ. Chang.* **2019**, *54*, 64–77. [CrossRef]

61. Laplaza, A.; Tanaya, I.G.L.P. Suwardji Adaptive comanagement in developing world contexts: A systematic review of adaptive comanagement in Nusa Tenggara Barat, Indonesia. *Clim. Risk Manag.* **2017**, *17*, 64–77. [CrossRef]

62. Irwansyah What do scientists say on climate change? A study of Indonesian newspapers. *Pacific Sci. Rev. A Nat. Sci. Eng.* **2016**, *2*, 58–65. [CrossRef]

63. Bohensky, E.L.; Kirono, D.G.C.; Butler, J.R.A.; Rochester, W.; Habibi, P.; Handayani, T.; Yanuartati, Y. Climate knowledge cultures: Stakeholder perspectives on change and adaptation in Nusa Tenggara Barat, Indonesia. *Clim. Risk Manag.* **2015**, *12*, 17–31. [CrossRef]

64. Ministry of National Development Planning/National Development Planning Agency (BAPPENAS). *National Action Plan For Climate Change Adaptation (RAN-API)*; Ministry of National Development Planning/National Development Planning Agency (BAPPENAS): Jakarta, Indonesia, 2012; p. 57.

65. Deressa, T.T.; Hassan, R.M.; Ringler, C.; Alemu, T.; Yesuf, M. Determinants of farmers' choice of adaptation methods to climate change in the Nile Basin of Ethiopia. *Glob. Environ. Chang.* **2009**, *19*, 248–255. [CrossRef]

66. Niles, M.T.; Mueller, N.D. Farmer perceptions of climate change: Associations with observed temperature and precipitation trends, irrigation, and climate beliefs. *Glob. Environ. Chang.* **2016**, *39*, 133–142. [CrossRef]

67. Mustafa, G.; Latif, I.A.; Bashir, M.K.; Shamsudin, M.N.; Daud, W.M.N.W. Determinants of farmers' awareness of climate change. *Appl. Environ. Educ. Commun.* **2018**, *0*, 1–15. [CrossRef]

68. Perez, C.; Jones, E.M.; Kristjanson, P.; Cramer, L.; Thornton, P.K.; Förch, W.; Barahona, C. How resilient are farming households and communities to a changing climate in Africa? A-based perspective. *Glob. Environ. Chang.* **2015**, *34*, 95–107. [CrossRef]

69. Glemarec, Y.; Qayum, S.; Olshanskaya, M. *Leveraging Co-Benefits Between Equality and Climate Action for Sustainable Development*; UN WOMEN: Songdo, South Korea, 2016; p. 55.

70. Quan, J.; Dyer, N. *Climate Change and Land Tenure: The Implications of Climate Change for Land Tenure and Land Policy*; Land Tenure Working Paper: London, UK, 2008; p. 62.

71. Mitchel, D.; McEvoy, D.; Antonio, D. A Global Review Of Land Tenure, Climate Vulnerability And A Capacity. In *Annual World Bank Conference on Land and Poverty*; The World Bank: Washington, DC, USA, 2018; p. 18.

72. FAO. *Adapting to climate change through land and water management in Eastern Africa: Results of pilot projects in Ethiopia, Kenya and Tanzania*; Food and Agricultural Organization: Rome, Italy, 2014; ISBN 978-92-5-108354-3.

73. Turral, H.; Burke, J.; Faures, J.-M. *Climate Change, Water and Food Security*; Food and Agricultural Organization: Rome, Italy, 2011; ISBN 978-92-5-106795-6.

74. DuBois, K.M.; Chen, Z.; Kanamaru, H.; Seeberg-Elverfeldt, C. *Incorporating Climate Change Considerations into Agricultural Investment Orogrammes: A Guidance Document*; Food and Agricultural Organization: Rome, Italy, 2012; p. 137.

75. Liu, C.L.C.; Kuchma, O.; Krutovsky, K.V. Mixed-species versus monocultures in plantation forestry: Development, benefits, ecosystem services and perspectives for the future. *Glob. Ecol. Conserv.* **2018**, *15*, 1–13. [CrossRef]

76. Lychuk, T.E.; Moulin, A.P.; Lemke, R.L.; Izaurralde, R.C.; Johnson, E.N.; Olfert, O.O.; Brandt, S.A. Climate change, agricultural inputs, cropping diversity, and environment affect soil carbon and respiration: A case study in Saskatchewan, Canada. *Geoderma* **2019**, *337*, 664–678. [CrossRef]

77. Stuart, D.; Schewe, R.L.; Mcdermott, M. Reducing nitrogen fertilizer application as a climate change mitigation strategy: Understanding farmer decision-making and potential barriers to change in the US. *Land Use Policy* **2014**, *36*, 210–218. [CrossRef]

78. Abdul-Rahaman, A.; Abdulai, A. Do farmer groups impact on farm yield and efficiency of smallholder farmers? Evidence from rice farmers in northern Ghana. *Food Policy* **2018**, *81*, 95–105. [CrossRef]

79. Liverpool-Tasie, L.S.O. Farmer groups and input access: When membership is not enough. *Food Policy* **2014**, *46*, 37–49. [CrossRef]

80. Ragasa, C.; Mazunda, J. The impact of agricultural extension services in the context of a heavily subsidized input system: The case of Malawi. *World Dev.* **2018**, *105*, 25–47. [CrossRef]

81. Davis, K.; Nkonya, E.; Kato, E.; Mekonnen, D.A.; Odendo, M.; Miiro, R.; Nkuba, J. Impact of Farmer Field Schools on Agricultural Productivity and Poverty in East Africa. *World Dev.* **2012**, *40*, 402–413. [CrossRef]

82. Morton, L.W.; McGuire, J.M.; Cast, A.D. A good farmer pays attention to the weather. *Clim. Risk Manag.* **2017**, *15*, 18–31. [CrossRef]

83. Liddell, T.M.; Kruschke, J.K. Analyzing ordinal data with metric models: What could possibly go wrong? *J. Exp. Soc. Psychol.* **2018**, *79*, 328–348. [CrossRef]

84. Ajuang, C.O.; Abuom, P.O.; Bosire, E.K.; Dida, G.O.; Anyona, D.N. Determinants of climate change awareness level in upper Nyakach Division, Kisumu County, Kenya. *Springerplus* **2016**, *5*. [CrossRef]

85. Mavisakalyan, A.; Tarverdi, Y. and climate change: Do female parliamentarians make difference? *Eur. J. Polit. Econ.* **2019**, *56*, 151–164. [CrossRef]

86. Gilligan, C. In a Different Voice: Psychological Theory and Women's Development. Available online: https://www.jstor.org/stable/20708485 (accessed on 6 April 2019).

87. Fosu-Mensah, B.Y.; Vlek, P.L.G.; MacCarthy, D.S. Farmers' perception and adaptation to climate change: A case study of Sekyedumase district in Ghana. *Environ. Dev. Sustain.* **2012**, *14*, 495–505. [CrossRef]

88. Abid, M.; Schneider, U.A.; Scheffran, J. Adaptation to climate change and its impacts on food productivity and crop income: Perspectives of farmers in rural Pakistan. *J. Rural Stud.* **2016**, *47*, 254–266. [CrossRef]

89. Runsten, L.; Tapio-Biström, M.-L. *Land Tenure, Climate Change Mitigation and Agriculture*; Food and Agricultural Organization: Rome, Italy, 2011; p. 2.

90. Fahad, S.; Wang, J. Farmers' risk perception, vulnerability, and adaptation to climate change in rural Pakistan. *Land Use Policy* **2018**, *79*, 301–309. [CrossRef]

91. Rondhi, M.; Adi, A.H. The Effects of Land Ownership on Production, Labor Allocation, and Rice Farming Efficiency. *Agrar. J. Agribus. Rural Dev. Res.* **2018**, *4*, 101–109. [CrossRef]

92. Rondhi, M.; Pratiwi, P.A.; Handini, V.T.; Sunartomo, A.F.; Budiman, S.A. Agricultural land conversion, land economic value, and sustainable agriculture: A case study in East Java, Indonesia. *Land* **2018**, *7*, 148. [CrossRef]

93. Rondhi, M.; Pratiwi, P.A.; Handini, V.T.; Sunartomo, A.F.; Budiman, S.A. Data on agricultural and nonagricultural land use in peri-urban and rural area. *Data Br.* **2019**, *23*, 103804. [CrossRef]

94. Lebel, L.; Käkönen, M.; Dany, V.; Lebel, P.; Thuon, T.; Voladet, S. The framing and governance of climate change adaptation projects in Lao PDR and Cambodia. *Int. Environ. Agreements Polit. Law Econ.* **2018**, *18*, 429–446. [CrossRef]

95. Lebel, L.; Sreymom, S.; Sokhem, P.; Channimol, K. Empirical and Theoretical Review of Climate Change and Water Governance to Enable Resilient Local Social-Ecological Systems. In *Climate Change and Water Governance in Cambodia: Challenges and Perspectives for Water Security and Climate Change in Selected Catchments, Cambodia*; Sam, S., Pech, S., Eds.; CDRI: Phnom Penh, Cambodia, 2015; ISBN 978-9924–500–04-9.

96. Ruishalme, I. Monoculture: Do Intensive Farming and GMOs Really Threaten Biodiversity? Available online: https://geneticliteracyproject.org/2017/07/14/monoculture-intensive-farming-gmos-really-threaten-biodiversity/ (accessed on 11 March 2019).

97. Shikuku, K.M.; Winowiecki, L.; Twyman, J.; Eitzinger, A.; Perez, J.G.; Mwongera, C.; Läderach, P. Smallholder farmers' attitudes and determinants of adaptation to climate risks in East Africa. *Clim. Risk Manag.* **2017**, *16*, 234–245. [CrossRef]

98. Nakano, Y.; Tsusaka, T.W.; Aida, T.; Pede, V.O. Is farmer-to-farmer extension effective? The impact of training on technology adoption and rice farming productivity in Tanzania. *World Dev.* **2018**, *105*, 336–351. [CrossRef]

99. Zhang, Q.; Yu, H.; Sun, P.; Singh, V.P.; Shi, P. Multisource data based agricultural drought monitoring and agricultural loss in China. *Glob. Planet. Chang.* **2019**, *172*, 298–306. [CrossRef]

100. Zhang, Q.; Gu, X.; Singh, V.P.; Kong, D.; Chen, X. Spatiotemporal behavior of floods and droughts and their impacts on agriculture in China. *Glob. Planet. Chang.* **2015**, *131*, 63–72. [CrossRef]

101. Kabir, M.J.; Alauddin, M.; Crimp, S. Farm-level adaptation to climate change in Western Bangladesh: An analysis of adaptation dynamics, profitability and risks. *Land Use Policy* **2017**, *64*, 212–224. [CrossRef]

102. Kahsay, G.A.; Hansen, L.G. The effect of climate change and adaptation policy on agricultural production in Eastern Africa. *Ecol. Econ.* **2016**, *121*, 54–64. [CrossRef]

 *land*

Article

Ecosystem Productivity and Water Stress in Tropical East Africa: A Case Study of the 2010–2011 Drought

Eugene S. Robinson [1], Xi Yang [2] and Jung-Eun Lee [1],*

[1] Department of Earth, Environmental, and Planetary Sciences, Brown University, Providence, RI 02912, USA; eugene_robinson@alumni.brown.edu

[2] Department of Environmental Sciences, University of Virginia, Charlottesville, VA 22904, USA; xiyang@virginia.edu

* Correspondence: leeje@brown.edu; Tel.: +1-401-863-6465

Received: 25 January 2019; Accepted: 12 March 2019; Published: 22 March 2019

Abstract: Characterizing the spatiotemporal patterns of ecosystem responses to drought is important in understanding the impact of water stress on tropical ecosystems and projecting future land cover transitions in the East African tropics. Through the analysis of satellite measurements of solar-induced chlorophyll fluorescence (SIF) and the normalized difference vegetation index (NDVI), soil moisture, rainfall, and reanalysis data, here we characterize the 2010–2011 drought in tropical East Africa. The 2010–2011 drought included the consecutive failure of rainy seasons in October–November–December 2010 and March–April–May 2011 and extended further east and south compared with previous regional droughts. During 2010–2011, SIF, a proxy of ecosystem productivity, showed a concomitant decline (~32% lower gross primary productivity, or GPP, based on an empirical SIF–GPP relationship, as compared to the long-term average) with water stress, expressed by lower precipitation and soil moisture. Both SIF and NDVI showed a negative response to drought, and SIF captured the response to soil moisture with a lag of 16 days, even if it had lower spatial resolution and much smaller energy compared with NDVI, suggesting that SIF can also serve as an early indicator of drought in the future. This work demonstrates the unique characteristics of the 2010–2011 East African drought and the ability of SIF and NDVI to track the levels of water stress during the drought.

Keywords: solar-induced chlorophyll fluorescence; drought; photosynthesis; East Africa; water stress; NDVI

1. Introduction

Droughts impact not only ecosystem functions but also the well-being of affected human populations [1,2]. The temporal and geospatial responses to drought are varied, and they threaten regions that are less well-adapted to water stress. A better understanding of the relationship between photosynthesis and water stress has broad reaching implications for our understanding of the global carbon and hydrological cycles.

Roughly half of the global variability in terrestrial carbon cycling can be attributed to carbon dioxide fluxes in tropical Africa, but the impact of drought on these fluxes is still modeled with significant uncertainty [3]. Further research on continental carbon fluxes is limited by data availability and characterization of regional productivity responses to drought [4,5]. The future of ecosystem productivity in the tropics is limited by how well plants cope with water stress but the influence of seasonal water stress on productivity, particularly in Africa's tropical regions, has rarely been characterized [6,7].

African climate is characterized by summer rainfall in the northern and southern tropics with an equatorial bimodal regime in-between. Regional variations notwithstanding (Figure 1), rainfall in tropical East Africa is delivered during the boreal spring and fall (Figure 2), accompanying local solar insolation maxima. The rainy season from March to May, known as the "long rains" in Kenya, "Belg" in Ethiopia, and "Gu" in Somalia, includes the majority of annual regional rainfall. A second rainy season occurs during the boreal fall and is referred to as the "short rains" in Kenya, "Keremt" in Ethiopia, and "Deyr" in Somalia. For simplicity, Kenyan terminology is employed in this study along with March, April, May (MAM) and October, November, December (OND) to refer to respective rainy months. In the area of our interest (black boxes in Figure 1), annual mean precipitation is 439 mm/year; this is a dry region where the two growing seasons (Figure 3) are susceptible to even small decreases in precipitation.

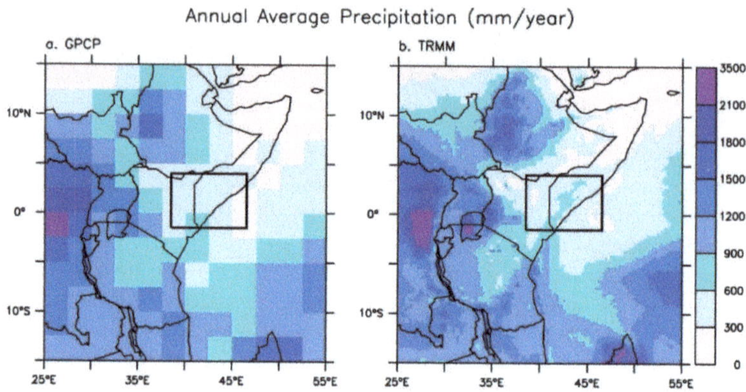

Figure 1. Average annual mean precipitation from Global Precipitation Climatology Project (GPCP; left) and Tropical Rainfall Measuring Mission Multisatellite (TRMM; right) climatology data, demonstrating the variability in rainfall regimes in the area. For zonally averaged values, the drought region is defined as the land area between 1.5° S and 4° N and 38.5° E to 46.5° E (black box).

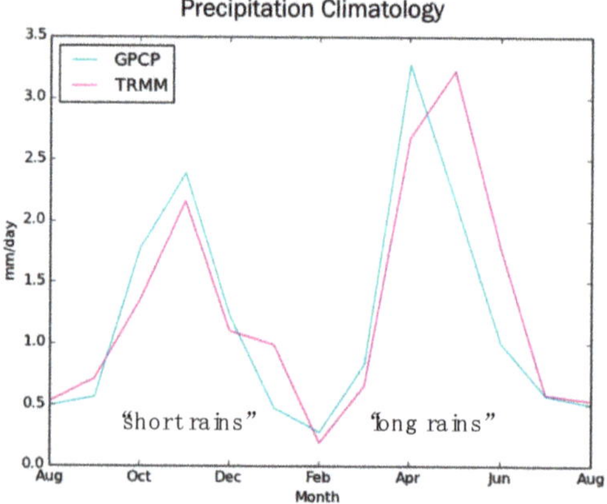

Figure 2. The climatology of rainfall over the drought region as presented by the two precipitation products used in this study, GPCP and TRMM.

Climate extremes on the African continent are well studied due to their potential to produce far-reaching economic and social consequences. East Africa has faced at least one major drought per decade over the last half century [8]. That said, regional droughts are typically isolated to only one of the two rainy seasons; consecutive failed rains are uncommon phenomena given the differing influences on precipitation for each rainy season. The successive failure of 2010 short rains and 2011 long rains produced the worst East African drought in the last 60 years [9], causing humanitarian crises in East Africa.

This study seeks to characterize the productivity response of tropical East Africa to water stress while identifying the unique spatial and temporal characteristics of the 2010–2011 drought. Related research on gross primary production (GPP) change in other tropical ecosystems and drought impacts relies on satellite-derived measurements because in situ measurements there remain scarce, largely due to operational constraints [10]. Here, we use relatively new satellite measurements of solar induced chlorophyll fluorescence (SIF) [11] to study the change in East African productivity during the drought, in conjunction with other atmospheric and terrestrial datasets. In Section 2, we describe the data and methods that we used. The results and discussion are presented in Sections 3 and 4. We summarize and conclude our study in Section 5.

2. Materials and Methods

2.1. Study Area

Our study area is the Horn of Africa region around the equator (Figure 1) where precipitation has two maxima (Figure 2). The Horn of Africa region, including Kenya, Ethiopia, and Somalia, collectively referred to as 'East Africa' (e.g., [12]), has been historically susceptible to droughts [13]. The El Niño Southern Oscillation (ENSO) significantly impacts OND precipitation [14]. The March and April precipitation signal, on the other hand, is more influenced by the position of the Intertropical Convergence Zone (ITCZ) (and thus the Indian Ocean warm pool temperatures [12]), and May rain is heavily influenced by the divergent low-level winds of the Indian monsoon [15].

2.2. Atmospheric Datasets

We used satellite-derived rainfall estimates from the Tropical Rainfall Measuring Mission Multisatellite (TRMM) Precipitation Analysis (TMPA) 3B43 product, averaged monthly from 1998 to 2013 at a 0.25-degree spatial resolution [16]. TRMM precipitation is calculated from the radiative and emissive properties of cloud hydrometers at visible, infrared, and microwave wavelengths, which serve as proxies for rainfall rate. The TRMM 3B43 dataset combines rain gauge, infrared, passive-microwave, and precipitation radar estimates and is generally well correlated at a monthly time scale with African rain gauge measurements across the continent [17] and over East Africa's complex topography [18]. To supplement TRMM data, we used the Global Precipitation Climatology Project (GPCP 1979–2012) [19] dataset because it has a longer record, starting from 1979, and is thereby valuable to compare the 2010–2011 drought with other historic droughts. GPCP data are derived from rain gauge and satellite data. We note that the number of rain gauges has declined throughout this region (40 in 1979 to 5 in 2010) [20].

Soil moisture data integrate precipitation anomalies in time and were used here to depict the spatial distribution of contemporary African droughts. Soil moisture products retrieved from Advanced Scatterometer (ASCAT) and Advanced Microwave Scanning Radiometer–Earth Observing System (AMSR–E) represents soil moisture in the top 1–2 cm of soil at a spatial resolution of ~50 km, sensitive to small precipitation events [21]. Land emissivity is a function of soil moisture; AMSR–E employs a low-frequency passive microwave remote sensing approach to measure the brightness temperature at Earth's surface [21]. The ASCAT is an active microwave sensor that measures backscatter from the surface, which is a function of soil moisture. This resulting soil moisture estimate demonstrates potential for drought monitoring [22]. Remotely-sensed soil moisture measurements are used in similar studies to monitor drought, including to characterize the spatial and temporal distribution of the 2010–2011 East Africa drought [23]. This study differentiates itself from previous research in that it characterizes the vegetation response to these conditions.

Reanalysis data from the European Centre for Medium-Range Weather Forecasts (ECMWF) were used to characterize the atmospheric and land surface conditions that led to the successive failure of the 2010–2011 East African monsoons [24]. Variables analyzed for drought monitoring extended from January 1979 to present and were analyzed to assess different drought effects and dynamics. Two m air temperature was selected due to its relationship to surface soil drying (as in [25]), its associated role as a climate driver (as in [26]), and because higher temperatures reduce water use efficiency during photosynthesis [27]; water vapor flux data were selected to monitor regional moisture transport. Sea surface temperature (SST) data were from the HadISST [28].

2.3. Terrestrial Datasets

Data related to both canopy structure and plant physiological activities were included in this study to characterize surface-level drought impacts. We used the normalized difference vegetation index (NDVI) from the moderate resolution imaging spectrometer (MODIS) MOD13C2 product. The resolution was 250 m, and we used data from 2007–2012 at 16-day, 0.05-degree resolution to characterize the change in potential or accumulated productivity [29]. We used solar induced chlorophyll fluorescence (SIF) as an indicator of actual photosynthetic activity from the Global Ozone Monitoring Experiment-2 (GOME-2)) [11]. GOME-2 measures this fluorescence signal at 9:30 am local time and provides global coverage every 1.5 days. The nadir footprint size is 40 km × 80 km; here we used both monthly and weekly GOME-2 products from NASA at a spatial resolution of 0.5°.

Fluorescence occurs when a solar photon is absorbed and is elevated to an excited state in the light reaction of photosynthesis. Typically, between 2 and 5 percent of photons absorbed by chlorophyll are re-emitted at longer wavelengths as fluorescence [30]. Canopy-level SIF measurements demonstrate that fluorescence capture productivity decreases even when NDVI remains constant [31]. SIF retrievals are also sensitive to seasonal dynamics of vegetation, independent from the structure of the canopy, and have been employed for stress detection in ground [32], aircraft [33], drone [34], and satellite-based instruments at wavelengths surrounding the oxygen A and B bands or Frauhoefer lines [35].

To assess the impact of the 2010–2011 drought on GPP, we scaled SIF to GPP using a linear relationship between monthly SIF and a GPP product, Fluxnet Multi-Tree Ensemble (MTE) GPP [36] during 2007–2011 (Figure 3). The Fluxnet GPP data were calculated using a machine-learning approach with eddy covariance datasets from flux towers, climatic variables, and remote sensing products. The dataset spanned from 1983 to 2011. The GOME-2 monthly SIF overlapped with Fluxnet–MTE over the 2007–2011 period. We then calculated the difference between 2010–2011 SIF and SIF climatology (2007–2012). We limited our analysis between 2007–2012 because GOME2 SIF showed a decreasing trend after 2013, likely caused by a sensor drift [37]. We calculated the GPP reduction during the drought period by using this relationship between SIF and GPP.

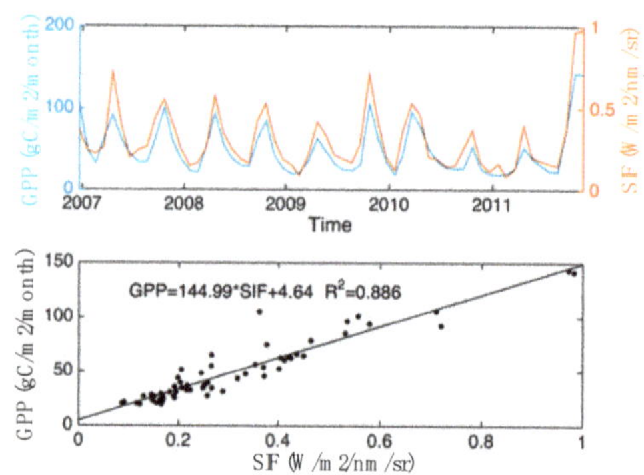

Figure 3. The relationship between solar-induced chlorophyll fluorescence (SIF) and gross primary production (GPP): (**top**) spatiotemporally-averaged SIF and GPP for the black box region in Figure 1; (**bottom**) the relationship between SIF and GPP.

2.4. Analysis

Remotely-sensed proxies for terrestrial vegetation activity (SIF and NDVI) and atmospheric reanalysis data were processed via the same methodology. These data were presented as standardized anomalies (z-scores), calculated by removing climatological mean values and dividing by the standard deviation of the mean. The resulting signal was therefore corrected for seasonal variability and the differences in the magnitude of the anomaly among different variables. In this analysis, multi-year means were calculated over as long a time period as the record permitted. For zonally averaged values, the drought focus region was defined as the land area between 1.5° S to 4° N and 38.5° E to 46.5° E (see black box in Figure 1), modified from [20].

In our analysis, the 2010–2011 drought was compared to previous droughts in East Africa as recorded in the emergency events database EM-DAT: The International Disaster Database, consistent with [38]. EM-DAT provides global historical drought records. In our study we included all EM-DAT-recorded drought events in Kenya, Ethiopia, and Somalia with the exception of 1987, 1988, and 1989, which were excluded because the drought season could not be determined. Previous regional drought years in the Horn of Africa include 1980 (MAM), 1983 (OND), 1991 (OND), 1994 (MAM), 1997 (MAM), 1998 (OND), 1999 (MAM), 2000 (MAM), 2003 (MAM), 2004 (MAM), 2005 (OND), 2008 (MAM), and 2009 (MAM). We note that 1998–1999 is the only other consecutive set of drought events before the 2010–2011 drought in the past 30 years.

3. Results

3.1. Spatial and Temporal Patterns of the 2010–2011 Drought

During the 2010–2011 drought, regional reductions in rainfall were evident spatially and temporally (Figures 4–6). The 2010–2011 drought, in fact, represented a significant reduction (as large as three standard deviations less) in rainfall, not only as compared to the climatology (Figures 5D and 6D), but also as compared to the region's substantial drought history (as large as 1.5 standard deviations; Figures 5F and 6F). During the previous drought years (a 'drought climatology' is presented in Figures 5B and 6B), a negative rainfall anomaly of the short rains (OND) was primarily between 40° E and 43° E, whereas the 2010 drought extended further east and with greatest intensity further westward than the average of 41.5° E (Figure 5E,F and Figure 6E,F). The decrease of the ensuing long rains (MAM) was one standard deviation below that of previous droughts (Figures 5B and 6B) in the southern hemisphere, exacerbating drought started in the season of short rains. In both rainy seasons, the 2010–2011 drought extended further east than in previous drought events. The 2010–2011 drought was spatially distinct from previous regional droughts.

The ITCZ does not extend over East Africa during the rainy seasons but its spatial coherence over the West Indian Ocean can be considered as a proxy for moisture transport to East Africa because the Indian Ocean temperature alters the local Walker Circulation [13]. East Indian Ocean SST is warmer than West Indian Ocean SST (Figure 7), and this east–west gradient in the SST pattern has been associated with droughts in East Africa [13]. A strong negative correlation exists between SSTs in the tropical Indian and Pacific Oceans and East African rainfall [39].

In August and September of 2010, the precipitation that eventually became the southern maximum of the West Indian Ocean double ITCZ was significantly reduced in intensity and showed a less robust spatial integrity than was evident in the climatology (Figures 6 and 7). Anomalously low precipitation over the western Indian Ocean in the months preceding the OND long rains yielded a reduced westward extension of the southern ITCZ and correspondingly reduced water vapor transport to East Africa. Ultimately, this resulted in less distinct double ITCZs over the West Indian Ocean and reduced precipitation over East Africa.

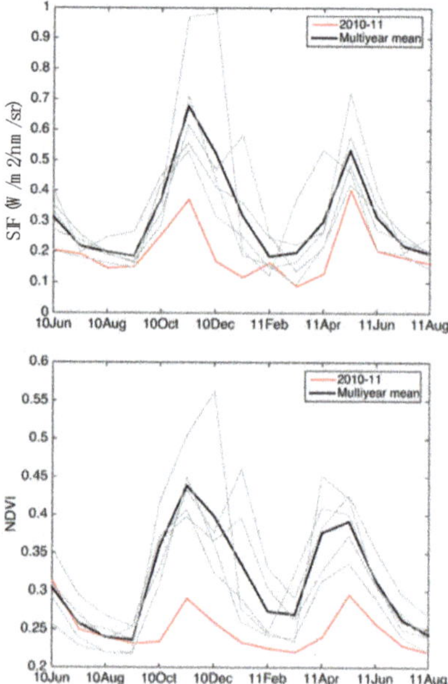

Figure 4. The 2010–2011 drought (red) represented a significant reduction of productivity (SIF and NDVI, Normalized Difference Vegetation Index) as compared to the mean annual cycle (thick black line) of different precipitation products averaged over the drought region.

Latitudinal and Temporal Precipitation Variation

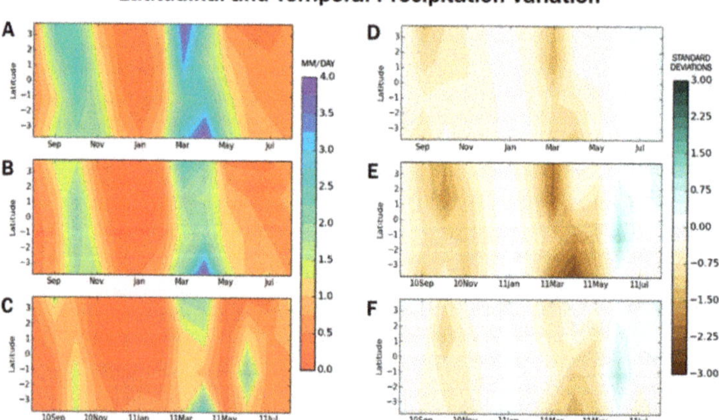

Figure 5. The 2010–2011 drought was spatially and temporally unique. The left column represents monthly rainfall climatology (GPCP) (**A**), drought climatology (alternatively referred to as a drought composite—an average of previous regional droughts (**B**), and 2010–2011 rainfall (**C**). Cool colors represent high rainfall. The right column represents average rainfall anomaly (B minus A) during drought years (**D**), 2010–2011 rainfall anomaly (C minus A) (**E**), and drought anomaly (E minus D or precipitation anomaly as compared to the drought composite average anomaly) (**F**). Brown values represent anomalously low rainfall whereas blue values represent greater than expected (anomalously high) rainfall. The anomaly was averaged along the longitude between 38.5° E and 46.5° E.

Longitudinal and Temporal Precipitation Variation

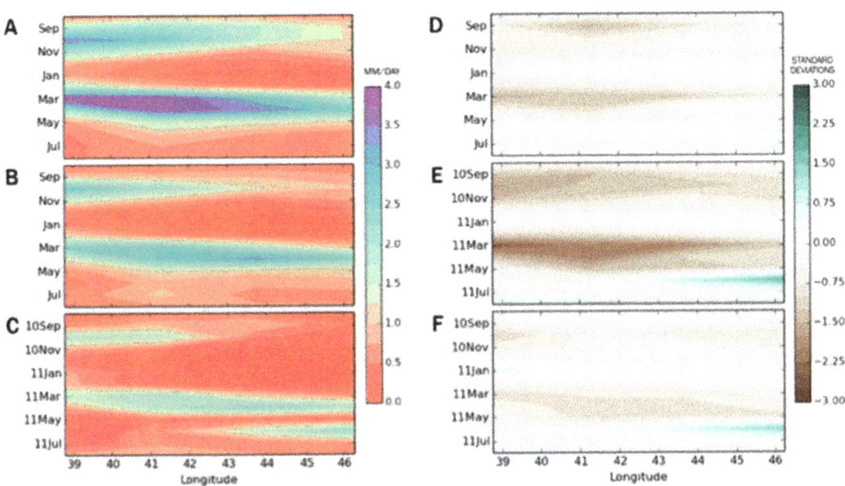

Figure 6. Similar to Figure 5, but the anomaly was averaged along the latitude between 1.5° S and 4° N.

Sea-Surface Temperature and Rainfall (TRMM): Climatology

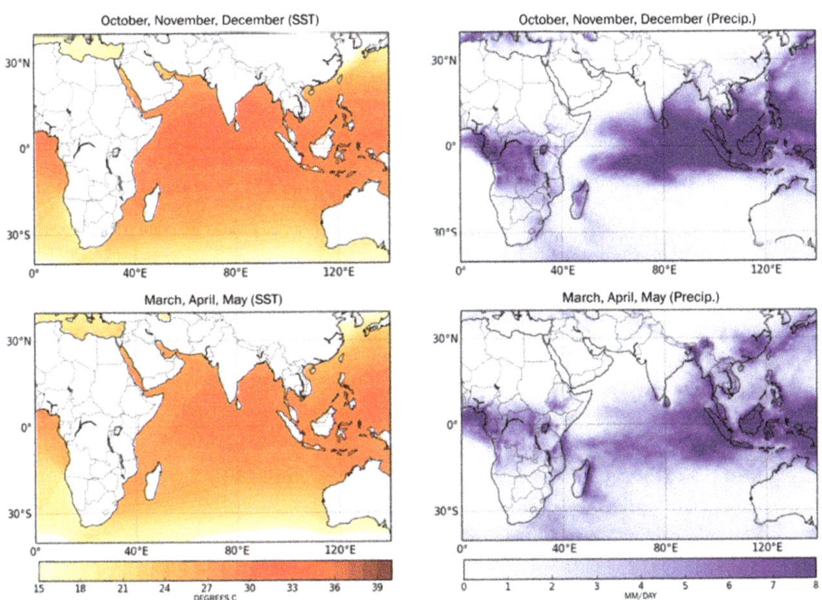

Figure 7. The climatology of sea surface temperature (HadISST, °C) and TRMM rainfall (mm/day) over the Indian Ocean during both rainy seasons. The East Indian Ocean is warmer than the West Indian Ocean. Moisture transport to East Africa is linked to this temperature gradient.

Failed formation of the double ITCZ over the West Indian Ocean continued into the MAM season, a season with a typically weaker double ITCZ (Figure 7). Precipitation in the West Indian Ocean was anomalously low during MAM of 2011, particularly over the southern maxima of the double ITCZ (Figure 8). Wind divergence from the southern ITCZ during January and February 2011, coupled with reduced long rains over the same area (Figure 9), produced a significantly weakened southern ITCZ, when compared to previous drought years. Water vapor for East African rainfall was heavily linked to this precipitation, and a relationship between rainfall in West Indian Ocean and East Africa was clear: strongly reduced rainfall in the West Indian Ocean resulted in (or was at least a strong proxy for) decreased precipitation in East Africa (e.g., [13]).

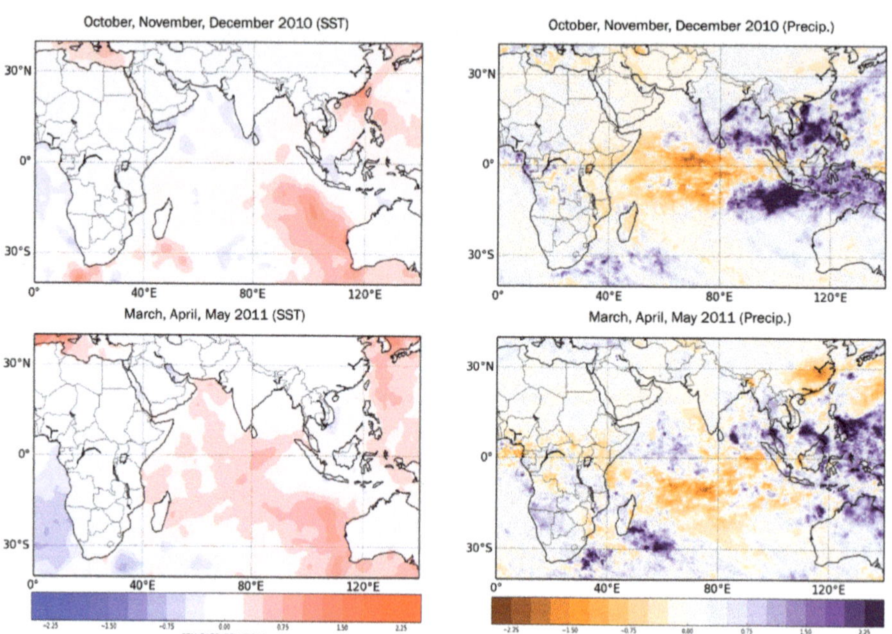

Figure 8. The anomaly of sea surface temperature (HadISST, °C) and rainfall (mm/day) during 2010–2011 East African drought.

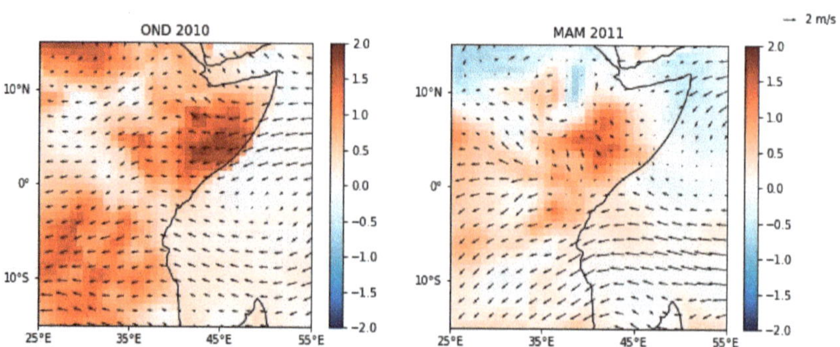

Figure 9. Anomalies of 2 m surface air temperature (filled colors) and 850 hPa wind (vectors). Temperatures generally increased in East Africa during the 2010–2011 failed rains. This contributes to but may also exist as a positive feedback with decreased photosynthetic activity: photosynthesis decreases at higher temperatures, but surface temperatures increase with the resulting reduced transpiration. OND is October, November, December; MAM is March, April, May.

3.2. Photosynthetic Responses to the Drought

SIF captures the spatial distribution and severity of the 2010–2011 East African drought (Figure 10). Anomalously low photosynthetic activity spatially correlated with soil moisture anomaly over the duration of the drought. MODIS NDVI spatially (Figure 10) and temporally (Figure 11) approximated the extent of the 2010–2011 drought, particularly within 10° of the equator.

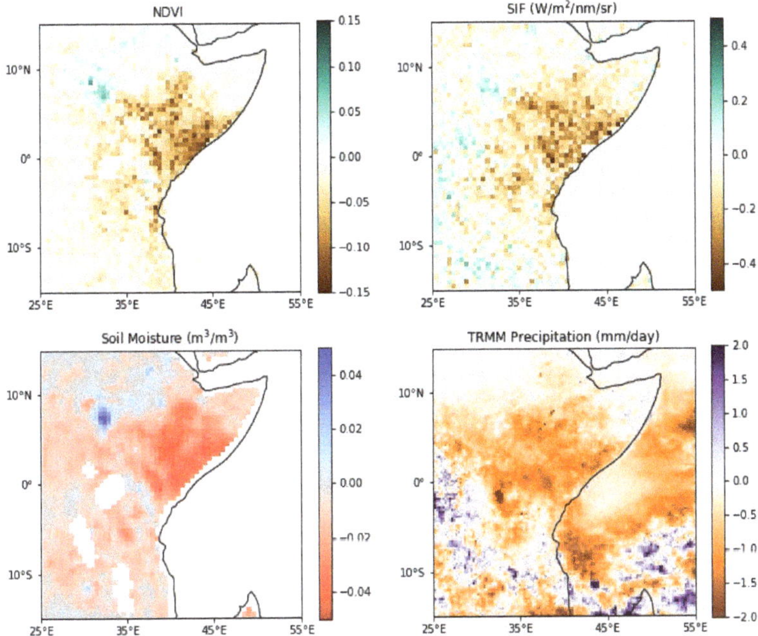

Figure 10. Anomalies of NDVI, SIF, soil moisture, and precipitation during the 2011 drought. Both SIF and NDVI approximate similar drought spatial extents. Anomalies calculated at highest available temporal resolution from the onset of the drought in October 2010 through April 2011.

The temporal response of NDVI or SIF followed soil moisture (Figure 11). The relationship was stronger with a lagged correlation (NDVI or SIF lags soil moisture) (Table 1), probably because the response of the ecosystem productivity took time. A 16-day lagged correlation has higher coefficient values (0.83 and 0.76 for NDVI and SIF) compared with no-lag correlation (0.56 and 0.35 for NDVI and SIF,). Both SIF and NDVI values were sensitive to soil moisture or water stress (Figure 11).

The small signal of SIF (1–2% of total APAR), and the large footprint of the GOME-2 sensor likely produced significant noise, as evidenced by the anomalies on the western extent of Figure 10. The considerably greater spatial resolution of the MODIS product likely yielded spatial precision in the MODIS NDVI panel of Figure 10.

As was demonstrated in previous studies [11,35], SIF scales well with GPP (Figure 3). We calculated GPP reduction using the relationship between GOME 2 SIF and Fluxnet–MTE GPP. For the long rain period, the total reduction in GPP was 117.47 gC m^{-2} (95% CI: 110.55, 124.40). For the short rain period, the total reduction in GPP was 64.4 gC m^2 (95% CI: 61.43, 67.38). The total reduction of GPP during this drought accounted for 32.1% of annual mean GPP (mean annual GPP in this area is 565.6 gC m^{-2}).

The spatial distribution of surface temperature anomalies calculated from ERAI monthly averages since 1979 suggested that decreased photosynthetic activity was correlated with temperature increases of between 0.8 and 1.3 degrees Celsius in the fall of 2010. Temperature anomalies were less pronounced in MAM 2011 (Figure 9).

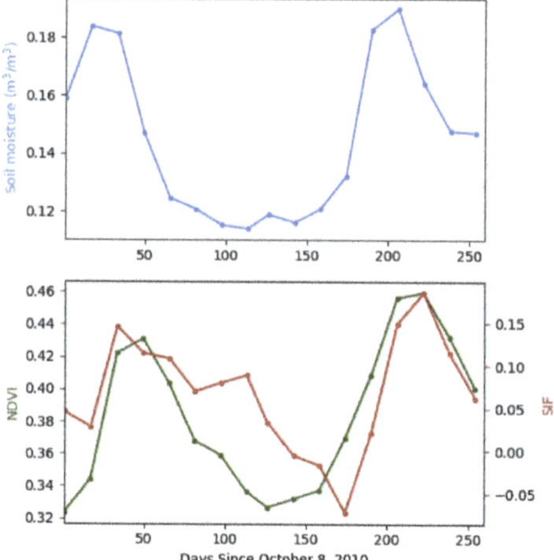

Figure 11. Zonally averaged SIF and NDVI response to soil moisture in East Africa (a black box in Figure 1. This figure was generated from 16-day average SIF from Global Ozone Monitoring Experiment-2 (GOME-2) data (red), NDVI from moderate resolution imaging spectrometer (MODIS) data (dark green), and daily soil moisture (blue) data.

Table 1. The relationship between soil moisture and NDVI or SIF from Figure 11. Lagged correlation is calculated as NDVI or SIF is 16 days lagging soil moisture.

	Correlation Coefficient	p
No lag		
Soil moisture and NDVI	0.56	0.02
Soil moisture and SIF	0.35	0.16
16-day lagged		
Soil moisture and NDVI	0.83	$p < 0.001$
Soil moisture and SIF	0.76	$p < 0.001$

4. Discussion

The mean annual cycle of precipitation in East Africa is already difficult to characterize in modeled results and interpolated data sets [40]. A dearth of regional climate data contributes to this problem: precipitation estimates are limited by a low density of rain gauges and the short record of high spatial resolution soil moisture data (SMOS data only dates back to 2010) [41]. Long term reductions to the long rains are linked to rising SSTs in the western Indian Ocean, a trend likely to continue as a consequence of anthropogenic climate change [12]. If the West Indian Ocean becomes warmer, long rains are projected to fail with increased frequency, as is evident since 1999.

Intergovernmental Panel on Climate Change (IPCC) projections for the region suggest a future characterized by increased episodic, extreme precipitation events, meaning there will be more rain and more droughts [42]. Other studies project an increase of rainfall in East Africa during the short rains in response to a large scale weakening of the Walker circulation [13]. Possible land-cover/land-use change as a consequence of agricultural technology adaption further exacerbates this uncertainty and are relevant to both natural ecosystems and agricultural regions, especially in food-insecure areas with high population densities [43].

Additionally, water stress is certain to have a substantial influence on natural ecosystems of East Africa in the coming decades. Previous research in the Amazon basin documented the impacts of drought on tropical ecosystems and was partially corroborated by this study [27]. Decreased water availability causes plant stomata to close, effectively shutting down photosynthesis. Increased temperature reduces the enzyme activity that enhances photosynthesis. Temperatures in East Africa are projected to rise with global warming and therefore the positive temperature anomaly during the OND 2010 drought offers insight into the productivity response of East Africa to this inevitable trend. Rising temperatures increase potential evaporation, thereby decreasing effective moisture. This mechanism is correlated with the drought stress influencing regional photosynthesis and suggests that water stress exhibits a first order control on ecosystem productivity in East Africa. A temporal analysis of surface temperature would be needed to corroborate this hypothesis and future work decoupling the relationship between temperature, water stress, and regional productivity is critical to an accurate projection of East Africa's future amidst climate change.

Previous research established the impact of rising global sea surface temperatures on the short rains: the dominant mode of variability in the tropical warm pool is related to global temperatures [39]—increased temperatures are likely to further dry East Africa due to the strong relationship between "short rains" and Indian Ocean sea surface temperatures. Efforts to link East African spring rains with sea surface temperature indices in the Pacific and Indian Ocean [44] continue, thus far suggesting that MAM rains are most sensitive to January sea surface temperatures (SSTs).

Both SIF and MODIS NDVI spatially approximated the extent of the 2010–2011 drought as compared to soil moisture and previous analysis [45], and SIF captured the response almost as well as did the NDVI, even if it has lower spatial resolution and much smaller energy, suggesting that SIF can also serve as an early indicator of drought in the future. However, state-of-the-art retrievals from GOME-2 are still limited by the sensor's design; the satellite was intended to monitor ozone, not SIF. Further, GOME-2 measurements occur at 0930 local time, well before the midday insolation and accompanying light saturation that causes stomata to close and thus shut down photosynthesis; vegetation may not be fully stressed at the time of SIF measurements. Taken together, these caveats suggest that GOME-2 SIF data may have under-reported water stress during the 2010–2011 drought. Measurements by an optimized sensor at peak insolation could serve as an improved record of vegetation stress.

The spatial extent of the failed 2011 long rains was amplified by the reduced availability of water resources following the failed short rains in the previous season. Failed rainy seasons amplify the consequences of subsequent reductions in rainfall. The converse, however, was not true in 2010–2011: anomalously high rainfall in MAM 2010 did not protect the region from the failed rains of OND 2010.

5. Conclusions

The failures of 2010 short rains (rainfall in OND) and subsequent long rains (rainfall in MAM) in East Africa caused a pronounced decrease in productivity, leading to a humanitarian crisis. Here we characterize the atmospheric and terrestrial response of tropical East Africa to the 2010–2011 drought. The 2010–2011 drought was extraordinary in temporal, spatial, and intensity anomalies even as compared to previous drought events. A stronger understanding of drought dynamics in East Africa will contribute to improved drought predictions [46] with hopes that their impacts can be prepared for in advance.

The successive failure of the rainy seasons in the 2010–2011 drought resulted in a devastating humanitarian crisis in East Africa. The decreasing trend of the long rains [12], particularly when the short rains fail as a result of the ENSO activity, could yield land cover changes and biome redistribution while fomenting social unrest and food insecurity. Future work related to the 2010–2011 East Africa drought will focus on determining the spatial variability of previous regional droughts to contextualize the unique features of the 2010–2011 event. An improved understanding of the ecosystem productivity response to water stress in these regions is critical due to the carbon impacts of increased drought incidence as a consequence of climate change. In the context of global changes in precipitation seasonality [47], this analysis concludes that SIF is an effective tool to track the response of vegetation to water stress. SIF responds to the water stress of the 2010–2011 East African drought, and it captures the response almost as well as the NDVI even if it has lower spatial resolution and much smaller energy.

Author Contributions: Conceptualization, J.-E.L.; Methodology, E.S.R., X.Y., J.-E.L.; Formal Analysis, E.S.R., X.Y.; Writing—Original Draft Preparation, E.S.R.; Writing—Review & Editing, E.S.R., X.Y., J.-E.L.; Visualization, E.S.R., X.Y.

Funding: This research received no external funding.

Acknowledgments: ESR thanks Greg Jordan-Detamore for his help with layout and design of figures. All data used in this study can be accessed online (see methods section).

Conflicts of Interest: The authors declare no conflict of interest.

References

1. Allen, C.D.; Macalady, A.K.; Chenchouni, H.; Bachelet, D.; McDowell, N.; Vennetier, M.; Kitzberger, T.; Rigling, A.; Breshears, D.D.; Hogg, E.H.; et al. A global overview of drought and heat-induced tree mortality reveals emerging climate change risks for forests. *For. Ecol. Manag.* **2010**, *259*, 660–684. [CrossRef]

2. Lesk, C.; Rowhani, P.; Ramankutty, N. Influence of extreme weather disasters on global crop production. *Nature* **2016**, *529*, 84–87. [CrossRef]

3. Fisher, J.B.; Sikka, M.; Sitch, S.; Ciais, P.; Poulter, B.; Galbraith, D.; Lee, J.E.; Huntingford, C.; Viovy, N.; Zeng, N.; et al. African tropical rainforest net carbon dioxide fluxes in the twentieth century. *Philos. Trans. R. Soc. Lond. B Biol. Sci.* **2013**, *368*, 20120376. [CrossRef]

4. Lewis, S.L. Tropical forests and the changing earth system. *Philos. Trans. R. Soc. Lond. B Biol. Sci.* **2006**, *361*, 195–210. [CrossRef]

5. Lewis, S.L.; Lopez-Gonzalez, G.; Sonke, B.; Affum-Baffoe, K.; Baker, T.R.; Ojo, L.O.; Phillips, O.L.; Reitsma, J.M.; White, L.; Comiskey, J.A.; et al. Increasing carbon storage in intact African tropical forests. *Nature* **2009**, *457*, 1003–1006. [CrossRef]

6. Guan, K.; Pan, M.; Li, H.; Wolf, A.; Wu, J.; Medvigy, D.; Caylor, K.K.; Sheffield, J.; Wood, E.F.; Malhi, Y.; et al. Photosynthetic seasonality of global tropical forests constrained by hydroclimate. *Nat. Geosci.* **2015**, *8*, 284–289. [CrossRef]

7. Schimel, D.; Stephens, B.B.; Fisher, J.B. Effect of increasing CO_2 on the terrestrial carbon cycle. *Proc. Natl. Acad. Sci. USA* **2015**, *112*, 436–441. [CrossRef]

8. Mwangi, E.; Wetterhall, F.; Dutra, E.; Di Giuseppe, F.; Pappenberger, F. Forecasting droughts in East Africa. *Hydrol. Earth Syst. Sci.* **2014**, *18*, 611–620. [CrossRef]

9. Loewenberg, S. Humanitarian response inadequate in Horn of Africa crisis. *Lancet* **2011**, *378*, 555–558. [CrossRef]

10. Anderegg, W.R.; Plavcova, L.; Anderegg, L.D.; Hacke, U.G.; Berry, J.A.; Field, C.B. Drought's legacy: Multiyear hydraulic deterioration underlies widespread aspen forest die-off and portends increased future risk. *Glob. Chang. Biol.* **2013**, *19*, 1188–1196. [CrossRef]

11. Frankenberg, C.; Fisher, J.B.; Worden, J.; Badgley, G.; Saatchi, S.S.; Lee, J.-E.; Toon, G.C.; Butz, A.; Jung, M.; Kuze, A.; et al. New global observations of the terrestrial carbon cycle from GOSAT: Patterns of plant fluorescence with gross primary productivity. *Geophys. Res. Lett.* **2011**, *38*. [CrossRef]

12. Lyon, B.; DeWitt, D.G. A recent and abrupt decline in the East African long rains. *Geophs. Res. Lett.* **2012**, *39*. [CrossRef]

13. Tierney, J.E.; Smerdon, J.E.; Anchukaitis, K.J.; Seager, R. Multidecadal variability in East African hydroclimate controlled by the Indian Ocean. *Nature* **2013**, *493*, 389–392. [CrossRef]

14. Nicholson, S.E.; Kim, J. The relationship of the El Niño—Southern oscillation to African rainfall. *Int. J. Climatol.* **1997**, *17*, 117–135. [CrossRef]

15. Camberlin, P.; Philippon, N. The East African March-May rainy season: Associated atmospheric dynamics and predictability over the 1968–97 period. *J. Clim.* **2002**, *15*, 1002–1019. [CrossRef]

16. Huffman, G.J.; Bolvin, D.T.; Nelkin, E.J.; Wolff, D.B.; Adler, R.F.; Gu, G.; Hong, Y.; Bowman, K.P.; Stocker, E.F. The TRMM Multisatellite Precipitation Analysis (TMPA): Quasi-Global, Multiyear, Combined-Sensor Precipitation Estimates at Fine Scales. *J. Hydrometeorol.* **2007**, *8*, 38–55. [CrossRef]

17. Herrmann, S.M.; Mohr, K.I. A Continental-Scale Classification of Rainfall Seasonality Regimes in Africa Based on Gridded Precipitation and Land Surface Temperature Products. *J. Appl. Meteorol. Climatol.* **2011**, *50*, 2504–2513. [CrossRef]

18. Dinku, T.; Ceccato, P.; Grover-Kopec, E.; Lemma, M.; Connor, S.J.; Ropelewski, C.F. Validation of satellite rainfall products over East Africa's complex topography. *Int. J. Remote Sens.* **2007**, *28*, 1503–1526. [CrossRef]

19. Adler, R.F.; Huffman, G.J.; Chang, A.; Ferraro, R.; Xie, P.P.; Janowiak, J.; Rudolf, B.; Schneider, U.; Curtis, S.; Bolvin, D.; et al. The version-2 global precipitation climatology project (GPCP) monthly precipitation analysis (1979-present). *J. Hydrometeorol.* **2003**, *4*, 1147–1167. [CrossRef]

20. Dutra, E.; Magnusson, L.; Wetterhall, F.; Cloke, H.L.; Balsamo, G.; Boussetta, S.; Pappenberger, F. The 2010–2011 drought in the Horn of Africa in ECMWF reanalysis and seasonal forecast products. *Int. J. Climatol.* **2013**, *33*, 1720–1729. [CrossRef]

21. Kolassa, J.; Gentine, P.; Prigent, C.; Aires, F. Soil moisture retrieval from AMSR-E and ASCAT microwave observation synergy. Part 1: Satellite data analysis. *Remote Sens. Environ.* **2016**, *173*, 1–14. [CrossRef]

22. Kolassa, J.; Gentine, P.; Prigent, C.; Aires, F.; Alemohammad, S.H. Soil moisture retrieval from AMSR-E and ASCAT microwave observation synergy. Part 2: Product evaluation. *Remote Sens. Environ.* **2017**, *195*, 202–217. [CrossRef]

23. Anderson, L.O.; Malhi, Y.; Aragao, L.E.; Ladle, R.; Arai, E.; Barbier, N.; Phillips, O. Remote sensing detection of droughts in Amazonian forest canopies. *New Phytol.* **2010**, *187*, 733–750. [CrossRef]

24. Dee, D.P.; Uppala, S.M.; Simmons, A.J.; Berrisford, P.; Poli, P.; Kobayashi, S.; Andrae, U.; Balmaseda, M.A.; Balsamo, G.; Bauer, P.; et al. The ERA-Interim reanalysis: Configuration and performance of the data assimilation system. *Q. J. R. Meteorol. Soc.* **2011**, *137*, 553–597. [CrossRef]

25. Zhou, L.; Tian, Y.; Myneni, R.B.; Ciais, P.; Saatchi, S.; Liu, Y.Y.; Piao, S.; Chen, H.; Vermote, E.F.; Song, C.; et al. Widespread decline of Congo rainforest greenness in the past decade. *Nature* **2014**, *509*, 86–90. [CrossRef]

26. Cook, K.H.; Vizy, E.K. Projected Changes in East African Rainy Seasons. *J. Clim.* **2013**, *26*, 5931–5948. [CrossRef]

27. Lee, J.E.; Frankenberg, C.; van der Tol, C.; Berry, J.A.; Guanter, L.; Boyce, C.K.; Fisher, J.B.; Morrow, E.; Worden, J.R.; Asefi, S.; et al. Forest productivity and water stress in Amazonia: Observations from GOSAT chlorophyll fluorescence. *Proc. Biol. Sci.* **2013**, *280*, 20130171. [CrossRef]

28. Rayner, N.A. Global analyses of sea surface temperature, sea ice, and night marine air temperature since the late nineteenth century. *J. Geophys. Res.* **2003**, *108*. [CrossRef]

29. Meroni, M.; Rossini, M.; Guanter, L.; Alonso, L., Rascher, U.; Colombo, R.; Moreno, J. Remote sensing of solar-induced chlorophyll fluorescence: Review of methods and applications. *Remote Sens. Environ.* **2009**, *113*, 2037–2051. [CrossRef]

30. van der Tol, C.; Berry, J.A.; Campbell, P.K.; Rascher, U. Models of fluorescence and photosynthesis for interpreting measurements of solar-induced chlorophyll fluorescence. *J. Geophys. Res. Biogeosci.* **2014**, *119*, 2312–2327. [CrossRef]

31. Daumard, F.; Champagne, S.; Fournier, A.; Goulas, Y.; Ounis, A.; Hanocq, J.-F.; Moya, I. A Field Platform for Continuous Measurement of Canopy Fluorescence. *IEEE Trans. Geosci. Remote Sens.* **2010**, *48*, 3358–3368. [CrossRef]

32. Yang, X.; Tang, J.; Mustard, J.F.; Lee, J.-E.; Rossini, M.; Joiner, J.; Munger, J.W.; Kornfeld, A.; Richardson, A.D. Solar-induced chlorophyll fluorescence that correlates with canopy photosynthesis on diurnal and seasonal scales in a temperate deciduous forest. *Geophys. Res. Lett.* **2015**, *42*, 2977–2987. [CrossRef]

33. Zarco-Tejada, P.J.; Morales, A.; Testi, L.; Villalobos, F.J. Spatio-temporal patterns of chlorophyll fluorescence and physiological and structural indices acquired from hyperspectral imagery as compared with carbon fluxes measured with eddy covariance. *Remote Sens. Environ.* **2013**, *133*, 102–115. [CrossRef]

34. Zarco-Tejada, P.J.; González-Dugo, V.; Berni, J.A.J. Fluorescence, temperature and narrow-band indices acquired from a UAV platform for water stress detection using a micro-hyperspectral imager and a thermal camera. *Remote Sens. Environ.* **2012**, *117*, 322–337. [CrossRef]

35. Guanter, L.; Zhang, Y.; Jung, M.; Joiner, J.; Voigt, M.; Berry, J.A.; Frankenberg, C.; Huete, A.R.; Zarco-Tejada, P.; Lee, J.E.; et al. Global and time-resolved monitoring of crop photosynthesis with chlorophyll fluorescence. *Proc. Natl. Acad. Sci. USA* **2014**, *111*, E1327–E1333. [CrossRef]

36. Jung, M.; Reichstein, M.; Margolis, H.A.; Cescatti, A.; Richardson, A.D.; Arain, M.A.; Arneth, A.; Bernhofer, C.; Bonal, D.; Chen, J.; et al. Global patterns of land-atmosphere fluxes of carbon dioxide, latent heat, and sensible heat derived from eddy covariance, satellite, and meteorological observations. *J. Geophys. Res.* **2011**, *116*. [CrossRef]

37. Zhang, Y.; Joana, J.; Gentine, P.; Zhou, S. Reduced solar-induced chlorophyll fluorescence from GOME-2 during Amazon drought caused by dataset artifacts. *Glob. Chang. Biol.* **2018**, *24*, 2229–2230. [CrossRef]

38. Masih, I.; Maskey, S.; Mussá, F.E.F.; Trambauer, P. A review of droughts on the African continent: A geospatial and long-term perspective. *Hydrol. Earth Syst. Sci.* **2014**, *18*, 3635–3649. [CrossRef]

39. Williams, A.P.; Funk, C. A westward extension of the warm pool leads to a westward extension of the Walker circulation, drying eastern Africa. *Clim. Dyn.* **2011**, *37*, 2417–2435. [CrossRef]

40. Mariotti, L.; Coppola, E.; Sylla, M.B.; Giorgi, F.; Piani, C. Regional climate model simulation of projected 21st century climate change over an all-Africa domain: Comparison analysis of nested and driving model results. *J. Geophys. Res.* **2011**, *116*. [CrossRef]

41. Sheffield, J.; Wood, E.F.; Chaney, N.; Guan, K.; Sadri, S.; Yuan, X.; Olang, L.; Amani, A.; Ali, A.; Demuth, S.; et al. A Drought Monitoring and Forecasting System for Sub-Sahara African Water Resources and Food Security. *Bull. Am. Meteorol. Soc.* **2014**, *95*, 861–882. [CrossRef]

42. Stocker, T.F.; Qin, D.; Plattner, G.K.; Tignor, M.; Allen, S.K.; Boschung, J.; Nauels, A.; Xia, Y.; Bex, V.; Midgley, P.M. *Cimate Change 2013: The Physical Science Basis. Intergovernmental Panel on Climate Change, Working Group I Contribution to the IPCC Fifth Assessment Report (AR5)*; Cambridge University Press: New York, NY, USA, 2013.

43. Moore, N.; Andresen, J.; Lofgren, B.; Pijanowski, B.; Kim, D.-Y. Projected Land-Cover Change Effects on East African Rainfall under Climate Change. *Int. J. Climatol.* **2015**, *35*, 1772–1783. [CrossRef]

44. Hoell, A.; Funk, C. Indo-Pacific sea surface temperature influences on failed consecutive rainy seasons over eastern Africa. *Clim. Dyn.* **2013**, *43*, 1645–1660. [CrossRef]

45. Hillbruner, C.; Moloney, G. When early warning is not enough—Lessons learned from the 2011 Somalia Famine. *Glob. Food Secur.* **2012**, *1*, 20–28. [CrossRef]

46. Funk, C.; Hoell, A.; Shukla, S.; Bladé, I.; Liebmann, B.; Roberts, J.B.; Robertson, F.R.; Husak, G. Predicting East African Spring Droughts Using Pacific and Indian Ocean Sea Surface Temperature Indices. *Hydrol. Earth Syst. Sci.* **2014**, *18*, 4965–4978. [CrossRef]

47. Feng, X.; Porporato, A.; Rodriguez-Iturbe, I. Changes in rainfall seasonality in the tropics. *Nat. Clim. Chang.* **2013**, *3*, 811–815. [CrossRef]

MDPI
St. Alban-Anlage 66
4052 Basel
Switzerland
Tel. +41 61 683 77 34
Fax +41 61 302 89 18
www.mdpi.com

Land Editorial Office
E-mail: land@mdpi.com
www.mdpi.com/journal/land